SpringerBriefs in Applied Sciences and Technology

PoliMI SpringerBriefs

Springer, in cooperation with Politecnico di Milano, publishes the PoliMI Springer-Briefs, concise summaries of cutting-edge research and practical applications across a wide spectrum of fields. Featuring compact volumes of 50 to 125 (150 as a maximum) pages, the series covers a range of contents from professional to academic in the following research areas carried out at Politecnico:

- Aerospace Engineering
- Bioengineering
- Electrical Engineering
- Energy and Nuclear Science and Technology
- Environmental and Infrastructure Engineering
- Industrial Chemistry and Chemical Engineering
- Information Technology
- Management, Economics and Industrial Engineering
- Materials Engineering
- Mathematical Models and Methods in Engineering
- Mechanical Engineering
- Structural Seismic and Geotechnical Engineering
- Built Environment and Construction Engineering
- Physics
- Design and Technologies
- Urban Planning, Design, and Policy

http://www.polimi.it

Umberto Tolino · Ilaria Mariani

Design Behind Interaction

Interfaces, Technology, Meanings

POLITECNICO
MILANO 1863

Springer

Umberto Tolino
Department of Design
Politecnico di Milano
Milan, Italy

Ilaria Mariani
Department of Design
Politecnico di Milano
Milan, Italy

ISSN 2191-530X ISSN 2191-5318 (electronic)
SpringerBriefs in Applied Sciences and Technology
ISSN 2282-2577 ISSN 2282-2585 (electronic)
PoliMI SpringerBriefs
ISBN 978-3-031-67415-0 ISBN 978-3-031-67416-7 (eBook)
https://doi.org/10.1007/978-3-031-67416-7

This Springer imprint is published by the registered company Springer Nature Switzerland AG
The registered company address is: Gewerbestrasse 11, 6330 Cham, Switzerland

Foreword

In our recent experience, we have witnessed a very interesting phenomenon, which is the growing interest in entrepreneurship. Today, unlike in past decades, many of the most successful companies (for example, in terms of capitalisation or brand awareness) are very young, born from the winning ideas of their founders and the intuition of those who believed in them and helped develop the idea and transform it into a business. Many of these successful companies arise from the combination of different sciences (medicine, biology, design, etc.) and technology. Therefore, I am very pleased to write the Preface to this book, which tells the comprehensive story of Thingk, a company that perfectly represents the "Politecnico mindset" to entrepreneurial development.

This book showcases the entrepreneurial journey of academic founders who have successfully harnessed the power of design-driven innovation to bridge the gap between scientific research and market opportunities. It demonstrates how Thingk, a university spin-off from Politecnico di Milano, rooted deeply in academic inquiry, has translated cutting-edge research into commercially viable products that redefine how we interact with technology daily.

This initiative is grounded in the thriving academic environment of Politecnico, where the fusion of technical knowledge and practical application was envisioned. The venture started from the reflection on how advanced research in smart technologies could be applied to create products that serve functional purposes but also integrate into the consumer lifestyle in new ways. It opens new avenues for market engagement and explores possibilities related to novel intuitive interactions embedded within everyday objects.

Central to the discussion is the strategic approach to technology transfer. The spin-off Thingk took advantage of the founders' ability to transform innovative ideas into solutions that answer existing and even anticipate emerging needs in the market. This process involved the identification of promising research directions and market gaps to envision product development opportunities, jointly adhering to academic rigour, market interests, and technological feasibility.

A significant aspect of the entrepreneurial approach taken has been the emphasis on applied experimentation, through an iterative process combining theoretical studies and emerging technologies into concepts to be translated into products. This approach allowed Thingk to bridge the gap between academic research and market application not only effectively, but in a surprising manner taking a different path than the majority of market products.

Each product developed by Thingk has both been a test of function and a market experiment at the same time. The product offerings were refined through direct engagement of prosumers or early adopters combined with a constant analysis of market trends. During the journey of this spin-off, strong emphasis has been placed on enhancing user experience through design innovation. This involves not just the aesthetics of products but their functionality and integration into the daily life of users. This iterative approach of actively engaging users and putting them at the centre of the development process led to products that are innovative from a technological standpoint and highly tailored to users' needs. Moreover, the applied experimentation triggered a deeper understanding of material properties and interface dynamics pushing the boundaries of traditional design paradigms. The experimentation ranged from novel approaches to user interaction with the help of embedded interfaces to exploring the potential of materials and materiality that are appealing and functional for users.

A distinctive approach concerns the constant attitude to engage with stakeholders and other innovators in the field of smart home appliances involving them in co-design activities.

The development of smart objects and interfaces of the highest quality has required a robust collaboration across various disciplines. The team includes a variety of experts among which designers, electronic- and computer science engineers, marketing experts and researchers with different humanistic backgrounds. The interdisciplinary collaboration has been crucial in developing products that are technologically advanced and providing a captivating user experience.

The result is a series of smart objects challenging traditional design paradigms and interaction patterns anticipating future market practices and technological adoption in the industry. This book outlines how such products were conceived laying out the backstory of their development.

The book outlines the constant commitment to innovation in the field of smart technologies with a design-driven approach exploring how innovation can spark when technology and design converge. With the ongoing evolvement of IoT and AI there are vast opportunities to expand the integration of technology into everyday objects, creating more intuitive and interactive user experiences.

This Preface introduces readers to the entrepreneurial vision behind the experimentation narrated in the book. A view that drives the blend of academic research and implemented, market-oriented solutions. It sets the stage for a deeper exploration of how theoretical knowledge can be effectively transformed into products introducing new ways of thinking our daily interaction with technology. The book seeks to inspire designers, engineers, and researchers to engage with and drive the frontier of technology and design-driven innovation, embracing an entrepreneurial vision.

Andrea Sianesi
Professor of Operations and Supply Chain Management
School of Management
Politecnico di Milano
Milan, Italy

Former President and Dean of MIP
the Business School of Politecnico di Milano (2015–2019)

Former President of Fondazione Politecnico di Milano (2020–2023)

Former President of PoliHub (2020–2024)

Preface

Digital transformation and technological innovations are fundamentally reshaping our perception and interaction with our surroundings, often prompting a revision of established design paradigms. This book explores how such advancements are affecting interactions, with experiences that unfold at the intersection of meanings, interfaces, and technology. It frames the current landscape, providing an overview of contemporary experimental approaches towards integrating interfaces into smart objects seamlessly. Some explorations even question the very concept of interfaces, moving towards their progressive disappearance and integration into products.

Within this context, the university spin-off Thingk engages in technology transfer, designing products that appear analog but are actually augmented with technology, incorporating multidisciplinary perspectives from product design, communication, and interaction design. The book examines Thingk's work in experimenting with embedded interfaces, spanning an eight-year journey of experimental design and analysis. It offers an in-depth exploration of how theoretical design research is applied in designing smart objects, leveraging context-awareness and situated meanings to the point of questioning and redefining conventional design concepts in terms of user interaction. Consequently, the significance of this book lies in its comprehensive analysis and insights into the design process behind such objects, underscoring the need for thorough examination of how semantic reconfigurations impact on affordances and agency.

Users play a crucial role in this discourse, from user research, co-creation, and crowdfunding to validation. Their engagement ensures that the development of smart objects and their interfaces is deeply informed by real-world application. Rooted in a research-through-design approach and relevant case discussions, this study aims to bridge the gap between theoretical exploration and practical application. Thus, it contributes to a deeper understanding of current experimentations with interfaces and smart technology, showcasing the rich potential of design-driven innovation in shaping future user experiences.

This book stands out in the research landscape for its in-depth exploration of embedded technology and interfaces within smart objects, leveraging a research-through-design methodology. It examines real-world case studies from the Thingk spin-off, demonstrating the application of theoretical research to market innovations. By adopting a multidisciplinary approach it provides a thorough understanding of how embedded technologies reshape user interactions, challenge traditional design paradigms and conventions, opening up relevant reflections on how hidden and variable interfaces cause semantic reconfiguration, the dynamic interplay of agency and affordances, but also the significant potentialities related.

Overall, this work is an attempt to mitigate the gap between theory and practical application, providing a bridge between academic inquiry and market realities.

Milan, Italy

Umberto Tolino
Ilaria Mariani

Acknowledgments

This book is the result of years of research and exploration, made possible through the invaluable support, expertise, and exceptional talent of many individuals.

We extend our deepest gratitude to all the co-founders of Thingk, especially Stefano Marangoni, whose constant support and knack for challenging our approaches with new projects have been a relevant booster. We are also thankful to Tommaso Livio, who has significantly enhanced our research activities with his keen sense of product aesthetics and interaction, and for his ability to identify fresh, innovative trends and directions to explore. Our thanks also go to Luca Venturelli for his efforts and collaboration in the *Slab!* project, helping us transition from crowdfunding to development and supporting the redesign and engineering of *Slab!*.

We are profoundly grateful to PoliHub, the innovation park and start-up accelerator at the Politecnico di Milano, and particularly to its former president, Andrea Sianesi, because PoliHub has been our nurturing home since the beginning, supporting us with training, funding opportunities, and networking opportunities within its vibrant community.

Our appreciation goes to the Politecnico di Milano, particularly the Department of Design and School of Design, for continuously feeding our research attitude and providing a fertile ground for innovation. The environment here has been foundational in encouraging our inquiries and explorations.

Lastly, this book owes much to those who collaborated with us and co-authored our past and ongoing works. For their outstanding support, we offer our heartfelt thanks to the significant people in our lives, including friends and family, who have been our constant backbone throughout this journey.

Contents

1 Introduction 1
1.1 Purpose and Significance 2
1.2 Background of This Book 3
1.3 Research Methodology 5
1.4 Aims and Impact 7
1.5 A Focus on the Expected Audience 8
References 10

2 Framing the Context Around Smart Objects. Between Interfaces and User Experiences 11
2.1 Setting the Stage 11
2.2 Observing Advanced Practices 14
2.2.1 Between Internet of Things and Everything 14
2.2.2 Artificial Intelligence of Things 15
2.2.3 Disappearing Interfaces 16
2.3 Reflecting on Gaps and Challenges 17
References 19

3 Technology Transfer. Bringing Scientific Research to the Market 23
3.1 Technology Transfer 23
3.1.1 Contribution of Science Parks and Incubator 24
3.1.2 Fundings and Academic Entrepreneurship 27
3.1.3 People and Organisational Culture 27
3.1.4 An Extended and Comprehensive Technology Transfer Ecosystem 28
3.1.5 The Role of the TTO 29
3.2 Product Development 30
3.2.1 Market Production 31
3.2.2 Research Prototypes 32
3.3 Patent Filing 35

3.4 Innovation Seeds 35
References 36

4 **A Pathway to Innovation. Design-Driven and User-Centric** 39
4.1 Design-Driven Innovation and a User-Centred Approach 39
4.2 The Design Process and Its Phases 41
4.2.1 Research 42
4.2.2 Design 44
4.2.3 Development 46
4.2.4 Slab!: The Process Applied 47
References 53

5 **Semantic Reconfigurations. Exploring Implications on Agency and Affordances** 55
5.1 Introduction: Making Sense of Things 55
5.1.1 Challenging Conventions by Design 56
5.2 Theoretical Framework 58
5.2.1 Domain of Reference 1: Variable Affordances 59
5.2.2 Domain of Reference 2: Embodied Interfaces 60
5.2.3 Domain of Reference 3: Embedded Technology 61
5.2.4 Domain of Reference 4: Expected Interaction 62
5.3 When Design is Misleading 65
5.4 Exploring Design Implications 67
5.4.1 Emphasising Shape-Function 68
5.4.2 Challenging Design Conventions 69
5.4.3 Semantic Reconfiguration 70
5.5 Design Issues: Discussing Challenges 73
References 76

6 **Shaping Hidden Interfaces for Situated Interactions** 79
6.1 Towards Context-Aware Interfaces 79
6.2 Drawing User Interfaces 81
6.2.1 (In)visibility and Control 81
6.2.2 Technology and Experimentation 83
6.2.3 Knob! 85
6.3 Perspectives: User Experience and Technology 87
References 88

7 **Conclusions and Future Research** 91
7.1 Making Sense of Things 91
7.2 Insights on Operationalising Theories 92
7.3 Main Contributions to Research 93
7.4 Limitations 94
7.5 Future Directions 94
References 96

Chapter 1
Introduction

Abstract This introductory chapter presents the core topic of the book, setting the stage for a detailed exploration of how embedded technologies are reshaping the way we interact with smart objects, focusing on the intersection of meanings, interfaces, and technology. The chapter outlines the book's main purpose: to investigate experimental approaches that challenge traditional interface concepts, advocating for their integration or even disappearance into objects to enhance user experiences. It highlights Thingk, a university spin-off, as a pivotal case study, demonstrating the practical application of theoretical research in the design of smart objects and IoT products.

Keywords Context of reference · University spin-off · Design methodology · Case study · Expected audience

This book is an essential exploration of the significant influence that digital transformation and technological advances are bringing, highlighting the transformative impact of embedded technology in design and human–computer interaction. Its primary purpose is to deep dive into the intricate interplay between such technological advancements and user experience design, offering a specific perspective on how objects and interfaces can be reimagined to enhance everyday human interactions. Specifically, it focuses on how embedded and emerging technologies are reshaping interactions at the intersection of meanings, interfaces, and technology itself. By framing the contemporary landscape and presenting significant case studies, the book goes into experimental approaches that question the very concept of interfaces, shifting towards their progressive disappearance and integration into objects.

U. Tolino and I. Mariani, *Design Behind Interaction*,
PoliMI SpringerBriefs, https://doi.org/10.1007/978-3-031-67416-7_1

1.1 Purpose and Significance

The book frames the contemporary landscape with a focus on Thingk, a university spin-off that exemplifies the transition of design theoretical research to market applications in the specific domain of smart objects and IoT products. By engaging in technology transfer activities and designing products that, while analog in appearance, are endowed with 'superpowers', Thingk conducts research and development in embedded technology and interfaces that challenge traditional concepts, leveraging user needs, context-awareness, and situated meanings. As such, the work of the spin-off symbolises an instance of digital transformation in design. This multidisciplinary work examines the implications these innovations have on established usage models and habits, as well as the impacts of technology in terms of agency and affordances. The book reports on insights from an eight-year journey of experimental design and analysis, linking theoretical knowledge with practical application through various case studies.

The significance of this work lies in its comprehensive analysis and practical insights into the development of smart, context-aware objects that challenge traditional design paradigms. It looks into current technological and design trends highlighting the need of thorough examination and understanding. This duality underscores the importance of both appreciating the experiential aspects of these trends and rigorously analysing their implications and developments.

Considering the significance of the topic and the variety of perspectives it includes, this work looks at current practices and state of the art in the field, tapping into the complex interplay between smart technology and the surrounding context, as well as possibilities and implications in terms of user experience. By delving into the concepts of agency, affordances, and semantic reconfiguration, it offers a novel framework for understanding the role of hidden interfaces and technology's potential to rethink user experiences. As such, it serves as a critical resource for understanding the multilayered dynamics of user interaction with embedded technologies, providing valuable guidelines for designing intuitive, user-friendly and innovative products. By examining objects that do not explicitly suggest their modes of interaction, it sheds light on the impacts of misleading or hidden interfaces on user perception and interaction. This focus is particularly relevant in discussions around how objects communicate their functionalities and how users navigate and interpret these innovative interfaces.

Significant emphasis is posed on the role of interfaces in shaping interactions, approaching them from a critical perspective. It views interfaces as key communicative devices and systems that convey multiple meanings, with the potential to significantly enhance the functionality and user experience of smart products. Delving into the development of variable interfaces that respond dynamically to user needs and environmental contexts, the book marks a departure from static design conventions towards a more fluid and responsive approach. This shift is crucial in understanding how interfaces can evolve to facilitate greater user engagement and richer interaction experiences.

Rather than focusing solely on generating foundational knowledge about smart objects and interfaces, this work is anchored in a specific epistemology derived from practice-based research. It seeks to enhance our understanding of practical applications and limitations, aiming to effectively bridge the gap between research and practical implementation in the evolving field of smart technology and interface design.

1.2 Background of This Book

The theoretical contribution draws from the core disciplines of product, communication, and interaction design, and extends to incorporate insights from adjacent fields. This includes cognitive psychology, which informs our understanding of user perception and decision-making processes; sociology and anthropology, providing context on the social and cultural implications of technology; and philosophy of technology, which offers a critical perspective on the broader impact of these advancements in society. This multidisciplinary approach enables a comprehensive exploration of smart objects and interfaces, anchoring our work in a rich array of scholarly perspectives.

Rooted in the discipline of design, which inherently revolves around the practice of creating, developing, and refining, this book adopts a holistic approach to the subject of smart objects and interfaces. It considers the entire spectrum of interaction design—from initial ideation to deployment, including iterative phases of testing and implementation. By oscillating between theoretical underpinnings and practical application, the book is structured as an amalgamation of conceptual frameworks and their pragmatic realisation.

The discourse is anchored in the methodology of research-through-design [9, 10, 15], a practice that involves designing, creating, and experimenting with artefacts as a primary means of enquiry. This approach consents to explore complex problems through the act of making, where the design process itself becomes a form of research. Moreover, it aligns with the notion of constructive design research [7, 9], which encourages the use of design artefacts as both a method of enquiry and ways for presenting research findings. This approach draws attention to the creation and testing of tangible artefacts as a means for nurturing and validating theoretical insights, thereby bridging the gap between abstract concepts and practical application. In the context of this research, research-through-design facilitates a hands-on exploration of smart objects and interfaces, treating them not just as products but as participants in a broader dialogue about technology, interaction, and human experience. This methodology resonates with our research, which revisits and discusses key themes explored in the spin-off, where smart objects and interfaces were investigated as dynamic agents of user interaction and experience. Through this lens, the design artefacts themselves become tools for inquiry, enabling a deeper understanding of their potential and implications in real-world contexts.

Moreover, the experimentation adopts a 'design as research' approach [2, 11, 16], which integrates the insights from existing scholarly work with reflections derived from our hands-on design practice. This book's contributions are shaped by a research-through-design approach [6, 8, 9], involving a series of studies conducted within national and international projects focused on the conception of smart products and novel interfaces.

By employing this method, our research not only contributes to the theoretical landscape but also provides practical, actionable knowledge directly applicable in the design of smart technologies.

Building on the experimentation at Thingk, this work extends from the processes of designing smart technology to understanding its impact on users. The chapters offer a comprehensive and timely contribution to the existing literature, which is often fragmented and predominantly focused on either case study analyses or theoretical explorations of smart object technology.

The content of this book is grounded in both empirical research conducted in the spin-off and foundational research within the Department of Design. This work benefits from a diverse range of experiments, studies, projects carried out over the years, which have contributed significantly to our knowledge. This process of continuous experimentation has provided practical insights and a deeper understanding of the field, enriching our approach with hands-on awareness and enhancing the overall discourse.

In addressing the current gaps identified through our applied research, the book navigates a path between theory and practice. Theoretical insights inform and enrich practical applications and vice versa, creating a symbiotic relationship between the two.

Previous investigations on which this work is grounded have stressed that the combination of designing and testing interfaces engaging end-users and relevant stakeholders are powerful sources of understanding and provide valuable insights [13, 18, 19]. These investigations have also shown that designing variable and dynamic interfaces introduces a variety of meanings and considerations into the design process. Building on these findings, this book adopts three complementary and transversal narrative lenses: technology, user experience, and aesthetics.

This book aims to uncover the multifaceted perspectives that render embedded technology and concealed interfaces compelling, intriguing, and adaptable for crafting innovative interactions with smart objects. While our findings and discussions may be applicable to embedded interfaces broadly, it is crucial to highlight that some reasonings are specific on a unique and challenging category of interfaces: those that are hidden and variable. We delve into interfaces that actively engage in meaning-making, either by inviting, suggesting, or challenging conventional usage. Our exploration includes a critical examination of designing interfaces that provoke a reevaluation of established habits.

1.3 Research Methodology

Acknowledging the current state of the art in the field, this book delves into the empirical research conducted within the Thingk spin-off and the Department of Design at Politecnico di Milano. It analyses the processes of both designing and testing smart objects mounting hidden interfaces. As anticipated, building upon eight years of comprehensive experimentation, it critically presents the findings and discusses the outcomes of eight years of experimentation across the entire process—from how the basic research fed experimentation to its testing and validation in co-creation sessions and impact on the overall design and development process.

The knowledge presented here is based on an array of activities and experimentations:

- The development of 7 products featuring hidden interfaces, of which 3 on the market and 4 research prototypes (one patented), alongside the concept of 'drawing interfaces'.
- The engagement with approximately 400 participants in crowdfunding (Indiegogo campaign) also engaged in co-creation for providing insights for refining product tested, especially working on aesthetics improvement that impacts the user experience.
- The engagement of about 100 persons in hands-on testing and validation conducted in workshop settings, employing focus groups and in-depth interviews to gather rich and comprehensive insights.
- The participation in national and international projects where the experimentation often implied a piloting role and opened up to in-depth technological experimentations, pushing further innovation.

Across all these activities, the primary goal is to explore the integration of novel interfaces within smart objects and assess their impact on user interactions and experiences. From a methodological point of view, these interfaces are examined as dynamic communication systems, capable of seamlessly transmitting a variety of information. Hence, the research approach spans from viewing design as an inquiry process to embracing experimentation as a means of knowledge acquisition. Through ethnographic analysis and interpretive research, we analysed our set of products, acknowledging the inherent limitations, biases, and weaknesses associated with each research method [1, 12, 14].

The research process behind this research and its methodology (summarised in Fig. 1.1) can be outlined as follows:

1. **Literature review:** Review of existing literature across the core and neighbouring disciplines to establish a comprehensive theoretical foundation, identify gaps in existing **knowledge, and situate the research within the broader academic discourse.**
2. **Analysis of products designed and the process behind their development:** Analysis of products and research prototypes that feature hidden and variable interfaces, such as drawing interfaces. This process translates theoretical insights

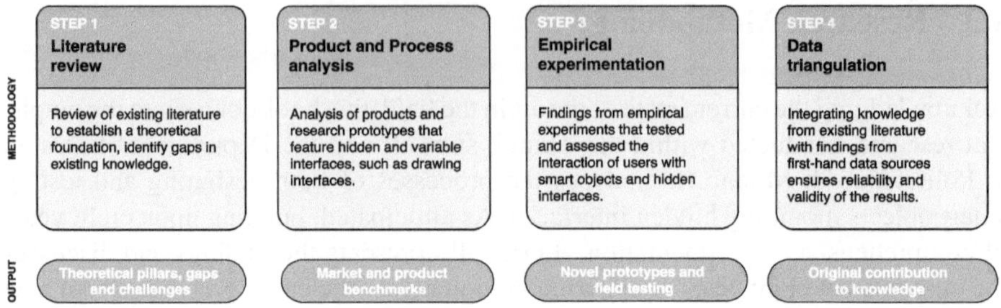

Fig. 1.1 Summary of the research methodology, highlighting steps, methodology, and output for each

into practical design solutions, with iterative refinements based on user feedback and observations.

3. **Empirical experimentation:** Findings from empirical experiments that tested and assessed the interaction of users with smart objects and hidden interfaces. This builds upon the designing, developing, and iteratively testing products to collect empirical data from end-users and stakeholders. The experimentation included organising targeted testing sessions, especially through workshops, complemented with focus groups, and in-depth interviews, with selected participants to validate the design choices and collect comprehensive feedback on user interaction and prototype usability. This often involves the engagement with broader communities through crowdfunding initiatives and/or co-creation activities to directly engage users and stakeholders in the development process. This allows for gathering a wide range of insights, preferences, and feedback to inform further iterations of the product design. In an alternative or complementary manner, the involvement in larger-scale research and innovation projects that offer opportunities for piloting the developed concepts and technologies allows for expanding the scope of experimentation, further pushing the boundaries of technological and design experimentation. Data is gathered employing a mixed-methods approach that combines qualitative and quantitative research techniques for comprehensive data collection, acknowledging that every research method has limits, biases and weaknesses [1, 3, 4, 14]. Beyond surveys and questionnaires, this includes participant observations[5, 17], focus groups, and interviews to deeply understand the context of use, user behaviours, and the social and cultural dimensions of interaction with smart objects.
4. **Data triangulation:** Integrating knowledge from existing literature with findings from first-hand data sources ensures reliability and validity of the results. This triangulation process not only validates the results against the current state of the art in the domain, but also offers a nuanced understanding of how Thingk provides original contribution to knowledge. This final stage involves analysing the implications of findings for both theory and practice, contributing to the field by bridging research and practical implementation.

These methodological stages form a comprehensive approach that combines theoretical exploration with practical experimentation, user engagement, and reflective analysis, ensuring a robust and insightful exploration of smart objects and interfaces.

Overall, the methodological approach adopted allows for exploring new technologies and assessing how innovative interfaces within smart objects influence interactions and diversify user experiences. By treating these interfaces as dynamic communication systems able to convey multiple information in a smooth and pleasant way. Ranging from examining *design* as a process of enquiry, to considering *experimentation* as a way to gain knowledge, we observed and studied our sample of products conducting ethnographic analysis and interpretive research.

1.4 Aims and Impact

This book aims to trigger the discourse and debate among peers, scholars, practitioners, and all those engaged with the context of smart objects, interfaces, and the transformative nature of embedded technologies. Originating from a rich blend of academic research and practical design experimentation within the domain of smart objects and interactive interfaces, this work is not merely a guidebook but an open dialogue on critical themes that emerged during our explorations. These themes, often underrepresented in the literature, contribute to gain a holistic understanding of the impact of digital transformation on design and human–computer interaction.

As both theorists and practitioners, we have recognised the need to account for pointing out attributes and potentials of smart technologies adopting a multidisciplinary lens. This recognition led us to identify key areas of inquiry that shaped the book's structure and content:

1. Technological foundations and interdisciplinary insights: How can we synthesise and build upon the diverse theoretical landscapes that inform smart object design?
2. Enhancing user experience: In what ways can designers craft interfaces that not only respond to user interactions but also enrich and transform the user experience?
3. Aesthetic considerations and semantic reconfiguration: How do aesthetics and the reconfiguration of meaning triggered by hidden and variable interfaces influence the design and perception of smart objects?
4. Hidden and variable interfaces: What are the challenges and opportunities in developing interfaces that adapt to their context and remain silent until engaged?
5. Impact on design practices and theory: How do these technological and design innovations inform and advance current design theories and practices?

The chapters of this book delve into these questions, offering a comprehensive examination of the evolving landscape of smart objects and interfaces, and their interplay:

- Chapter 2 sets the stage by discussing how smart objects are influencing our lives, framing them within a broader context of technology, interfaces, and user experience.
- Chapter 3 explores the journey of technological research into market-ready products, highlighting the role of innovation ecosystems in bridging the gap between theory and application.
- Chapter 4 presents a detailed look at the innovation process, from conceptual design to product development, using case studies like Slab! to illustrate the integration of complex functionalities within user-centric designs.
- Chapter 5 examines the challenges and implications of designing objects that communicate their functionality in non-traditional ways, impacting our understanding of agency and affordance in design.
- Chapter 6 focuses on the development and implications of variable interfaces that respond to user needs and environmental contexts, showcasing Thingk's experimental approaches and case studies.

Each chapter employs one or multiple case studies to bridge theory and practice, showing how theoretical knowledge is operationalized in various contexts.

1.5 A Focus on the Expected Audience

The insights presented in this book are crafted for a diverse audience, including researchers and academics in human–computer interaction and smart technologies, design researchers and practitioners focused on interactive and IoT product design, and entrepreneurs working on digital transformation aspects in the tech industry.

By offering a blend of theoretical exploration and practical case studies, which builds on the extensive experimentation within the university spin-off Thingk, this work contributes to knowledge by offering a comprehensive, empirically grounded, and innovative exploration of how digital transformation and embedded technology are reshaping design, human–computer interaction, and user experience.

It is particularly beneficial for those looking to understand the interplay between design theory and market application, exploring the innovative and often overlooked aspects of embedded technology in product design. The book challenges traditional paradigms by examining how subtle design choices in interaction modes can significantly impact user perception and experience. As such, it invites these diverse audiences to engage in a reflective journey, reevaluating their approaches to design, functionality communication, and the broader implications of their work in shaping user experience in the digital era.

From a design perspective, the book aims at affecting designers and researchers exploring interactions triggered by embedded technologies and interfaces, and related issues often neglected or overlooked. Stressing in particular the interplay between design theory and practice, as the transition of theoretical research to market applications, this work opens necessary reflections into semantic reconfiguration that embedding technology requires when designing smart objects, ranging from the design implications, communication of functionalities and usability. This section challenges traditional design paradigms, exploring objects that do not explicitly indicate their modes of interaction and how this influences perceptions of agency and affordances. As such, it invites reflection on the implications of design choices in such a challenging domain.

The insights and discussions presented throughout this book are particularly valuable for entrepreneurs in the field of design and technology. For these professionals, the book serves as a source of inspiration and innovative ideas. It looks into the complexities of smart object design, embedded technology, and user-interface interactions, offering suggestions, interpretations, and practical use cases that can spark creative and forward-thinking solutions. The book can be seen as an inspiration to effective and efficient design processes and approaches. Entrepreneurs, especially those in startups or small-scale operations, can benefit greatly from the in-depth exploration of interaction design and embedded technology. This knowledge can inform their product development strategies, aiding in the creation of technologically advanced yet user-friendly products. Importantly, the book impacts the way commercial products are conceptualised and developed. It offers insights into semantic reconfiguration and the influence of design on user perception and interaction, vital for creating intuitive and engaging products that resonate with users and stand out in the market. Finally, the book supports strategic decision-making by illustrating the crucial interplay between technology, design, and user experience. This understanding is essential for entrepreneurs aiming to meet and surpass market expectations, ensuring their products are not only technologically sophisticated but also align with modern consumer preferences.

Grounded in a comprehensive research-through-design approach, this book is built upon an eight-year journey of experimental design, research, and innovation. It showcases how the proposed advancements on interfaces not only transform the user experience with smart objects but also challenge established design paradigms, paving the way for future innovations in interface design.

* * *

This book is the result of eight years of research and experimentation at Thingk, involving the collective efforts of the entire team of the spin-off. Within this team, **Umberto Tolino** played a dual role as both a designer and a researcher, contributing his unique expertise in design methodologies and practical application. **Ilaria Mariani** primarily focused on academic research, taking charge of literature review, UX analysis, and data interpretation.

Both authors have contributed equally to this publication, from the conceptualisation to the writing, review and editing. In terms of drafts, Ilaria Mariani primarily contributed to Chaps. 2 and 4; Umberto Tolino to Chaps. 3 and 7; while Chaps. 1, 5, and 6 were fully drafted together. Then, throughout the design and writing process of the book, all chapters have been collaboratively edited and refined by both authors, ensuring a cohesive and comprehensive presentation of their collective insights and findings.

References

1. J.W. Creswell, *Research Design: Qualitative, Quantitative, and Mixed Methods Approaches* (Sage, Thousand Oaks, CA, 2008)
2. N. Cross, *Designerly Ways of Knowing* (Springer, London, UK, 2006). https://doi.org/10.1007/1-84628-301-9
3. N.K. Denzin, Y.S. Lincoln (eds.), *Handbook of Qualitative Research* (Sage, London, UK, 1994)
4. N.K. Denzin, Y.S. Lincoln, The discipline and practice of qualitative research. Handb. Qual. Res. **2**, 1–28 (2000)
5. K.M. DeWalt, B.R. DeWalt, *Participant Observation: A Guide for Fieldworkers* (AltaMira Press, Walnut Creek, CA, 2010)
6. A. Findeli, A quest for credibility: doctoral education and research in design at the University of Montreal. Doctoral Education in Design, 1 (1998)
7. P. Gall Krogh, Drifting by intention—four epistemic traditions in constructive design research. in *Extended Abstracts of the 2022 CHI Conference on Human Factors in Computing Systems* (Association for Computing Machinery, New York, NY, USA, 2022). https://doi.org/10.1145/3491101.3503756
8. W. Jonas, Design research and its meaning to the methodological development of the discipline, in *Design Research Now: Essays and Selected Projects*, ed. by R. Michel (Birkhäuser Basel, Basel, 2007), pp. 187–206. https://doi.org/10.1007/978-3-7643-8472-2_11
9. I. Koskinen, J. Zimmerman, T. Binder, J. Redstrom, S. Wensveen, *Design Research Through Practice: From the Lab, Field, and Showroom* (Elsevier, Burlington, MA, 2011)
10. P.G. Krogh, T. Markussen, A.L. Bang, Ways of drifting—five methods of experimentation in research through design, in *ICoRD'15—Research into Design Across Boundaries*, vol. 1, ed. by A. Chakrabarti (Springer India, New Delhi, India, 2015), pp. 39–50. https://doi.org/10.1007/978-81-322-2232-3_4
11. B. Laurel, *Design Research: Methods and Perspectives* (The MIT Press, Cambridge, MA, 2003)
12. Y.S. Lincoln, S.A. Lynham, E.G. Guba, Paradigmatic controversies, contradictions, and emerging confluences, revisited, in *The Sage Handbook of Qualitative Research*, vol. 4 (2011), pp. 97–128
13. I. Mariani, T. Livio, U. Tolino, Drawing Interfaces. When interaction becomes situated and variable, in *DeSForM 2019 Proceedings* (DeSForM, Boston, MA, 2019). https://desform19.pubpub.org/pub/drawing-interfaces/release/1
14. V.L. Plano Clark, J.W. Creswell, *The Mixed Methods Reader* (Sage, Thousand Oaks, CA, 2008)
15. D. Raptis, R.H. Jensen, J. Kjeldskov, M.B. Skov, Aesthetic, functional and conceptual provocation in research through design, in *Proceedings of the 2017 Conference on Designing Interactive Systems* (Association for Computing Machinery, New York, NY, USA, 2017), pp. 29–41. https://doi.org/10.1145/3064663.3064739
16. D.A. Schön, *The Reflective Practitioner: How Professionals Think in Action* (Basic Books, New York, NY, 1983)
17. R.E. Stake, Qualitative case studies, in *Handbook of Qualitative Research*. ed. by N.K. Denzin, L. Lincoln (Sage, London, UK, 1994), pp. 443–366
18. U. Tolino, I. Mariani, Do you think what i think? Strategic ways to design product-human conversation. Strat. Des. Res. J. **11**(3), 254–262 (2018). https://doi.org/10.4013/sdrj.2018.113.10
19. U. Tolino, I. Mariani, Hacking meanings. Innovation as everyday invention. DIID Disegno Ind. **65**(18), 78–85 (2019)

Chapter 2
Framing the Context Around Smart Objects. Between Interfaces and User Experiences

Abstract This chapter delves into the transformative role of smart objects in daily life, illustrating their ubiquity and intrinsic value. It establishes a comprehensive theoretical framework that links smart objects with technology, interfaces, and user experience. Finally, it discusses the benefits and challenges of embedding hidden and variable interfaces on everyday objects, considering how they affect user interaction in terms of usability, accessibility, and the overall relationship between humans and technology.

Keywords Smart objects · Human–computer interaction · User interfaces · IoT · AIoT

2.1 Setting the Stage

In today's technological landscape smart objects have become increasingly intrinsic, ubiquitous, and pervasive to our daily lives [30], reshaping our interactions with technology and each other [15, 18, 40]. As these objects seamlessly integrate into various aspects of our environment, there emerges a need to frame a comprehensive discourse around them. This discourse must not only consider the technological underpinnings of smart objects, but also delve into the features of their interfaces and the user experiences they facilitate [14, 27, 46]. To lay the groundwork for such a discussion, it is necessary to construct a theoretical framework that links smart objects with the broader concepts of technology, interfaces, and user experience [25, 34, 51].

Given the rapid succession of technological advancements, their impact extends beyond merely altering how we interact with our surroundings; they are reshaping it profoundly [7]. Technology is becoming increasingly ubiquitous and pervasive, as well as more human-centric, enabling, and accessible [30]. affects our relationships with others, with objects, and with the environment, prompting a spectrum of changes. These changes, whether subtle or overt, compel us to reconsider our interaction modes. This ongoing stream, ranging from behavioural adaptations to

U. Tolino and I. Mariani, *Design Behind Interaction*,
PoliMI SpringerBriefs, https://doi.org/10.1007/978-3-031-67416-7_2

acquiring new digital literacies, represents a significant cultural evolution. Consider the growing technological pervasiveness in everyday devices, which has transformed our environments to become smarter, more interactive, and better connected. This connectivity extends not only among devices themselves but also with their surrounding contexts and with users. Such advancements foster a myriad of innovative and sophisticated experiences, resulting in novel usage models and paradigms. These developments necessitate a reevaluation of the human-technology relationship, suggesting a shift towards more integrated and interactive environments [31]. In light of this, the fundamental goal extends beyond merely simplifying our lives and daily activities. It also addresses needs often created by technology itself [35], linking people and objects within ecosystems that are complex yet deeply interconnected.

Since the early 2000s, the field of Human–Computer Interaction (HCI) has experienced a pivotal shift from viewing people and computers as distinct entities to an integrated perspective that includes technology, intelligent materials, and humans [49, 50]. Particularly, the significant reduction in the cost of electronic components over the past few decades has fueled a new phase of electrification and digitalization, which has seen rapid development, especially in the domestic sphere [41]. Especially within the Smart Home Appliances sector [4], new interactions have emerged that are destined to improve, rethink, or replace existing experiences [3, 8, 12, 32, 39]. Following a principle that conceives the system as holistic and interconnected, we are witnessing a redesign of our environments which are becoming intelligent as they host distributed technologies [36, 51]. These technologies form networks of devices and objects equipped with sensors, actuators, and controllers that process and exchange information and act accordingly. This paradigm shift has facilitated dynamic, omnipresent, and sometimes nearly invisible interactions [29], aligning with the principles of calm technology [48, 52]. Embedded technology and the notion of embedded interactions [28] play a pivotal role in the discourse around smart objects, enabling technology to blend seamlessly into object functionality, thereby enhancing functionality while making interactions more intuitive.

The integration of computational capabilities into our lives and daily activities has reached such a level that it allows us to conceptualise technology itself as a design material that operates in symbiosis with other physical materials, within an integrated ecosystem enabling new experiences and usage practices. The goal is to achieve satisfying interactions without the need to provide explicit information, as these are detected and processed by intelligent machines capable of understanding, predicting, and interacting with us and other computationally powerful machines. The focus is on designing and envisioning artefacts that are sensitive to their surroundings, connected to the Internet and/or each other, governed by a code that regulates their behaviour, and thus defining how they interact with other objects, applications, and even environments. Incorporating computational power into objects and networking them equips them with intelligence and enables interactions. Considering and implementing these aspects outline a challenging field of inquiry requiring hybrid interdisciplinary approaches, where design meets and often merges with engineering.

As a consequence of this paradigm shift toward embedded technologies and calm technology principles, user interfaces have undergone significant transformation. Traditional interfaces that once demanded active engagement and attention are turning into more ambient, integrated forms. This change aligns with the shift from distinct, tangible interaction mechanisms toward more fluid and less intrusive modalities [16]. Consequently, interfaces are growingly designed to be minimally invasive yet highly functional, allowing the user to perform natural interactions with their environment without suffering the cognitive burden traditionally associated with technology use and the mediation it causes [37]. Applied to smart devices, this approach is translated into provision of necessary information and controls only when needed, thereby reducing overall user effort and enhancing the seamlessness of the human-technology interaction. These advancements signify a move towards an ecosystem where technology not only supports but also anticipates user needs, fostering an intuitive and almost unconscious interaction that aligns with our daily habits and sensory inputs.

Current technology acts as an enabler, fostering a symbiotic relationship with other smart objects, and creating an ecosystem that nurtures new experiences and usage practices [10, 11, 47]. A key trend is the evolution of these technologies to become more user-friendly [3, 4, 15]. In this sense, a further paradigmatic spur comes from the introduction of AI in the discourse. The convergence of the Internet of Things (IoT) and AI has birthed the concept of the Artificial Intelligence of Things (AIoT). In this advanced interdisciplinary field, IoT functions analogously to a nervous system that networks different elements, gathering data which the AI then processes for decision-making. As such, AIoT enriches IoT devices with its AI capabilities, enabling them to operate more effectively and adaptively, offering intelligent functions and autonomous learning using advanced AI techniques.

The integration of AI represents a profound evolution for IoT, which shifts from simple interconnectivity among devices to a more complex system capable of intelligent data analytics and enhanced operational efficiency. AIoT leverages data streams generated by IoT environments [1], employing AI to enable more sophisticated data processing directly at the network's edge. This innovative paradigm emphasises the fusion of real-time data capture via IoT sensors with advanced analytics to create an intelligent ecosystem. AIoT not only enhances the data processing capabilities of traditional IoT systems but also fosters an intelligent interconnectivity that spans devices, systems, and infrastructures [9]. This results in a fully digitalised and intelligently interconnected environment, enhancing both efficiency and effectiveness across various applications and industries. Through AIoT, devices not only communicate but also independently analyse and respond to environmental data, pushing the boundaries of what smart systems can achieve in real-time decision-making and automation.

In this context, the more complex task assigned to machines involves collecting data, analysing it, and acting accordingly, while the user focuses on overseeing or monitoring the process. The enhancement of computers and devices with systems that enable information gathering allows objects to interpret the surrounding world, moving beyond the limitation of merely receiving data directly from users [2].

Concurrently, the significant benefit of an increasing computational capacity (ubiquitous computing) lies in reducing the amount of information users need to process and remember during activities, addressing what Sweller refers to as cognitive load [45]. Building upon this, when AI integrates into the loop with ubiquitous computing, it adds an intelligent layer that enhances decision-making and predictive capabilities, hence going beyond traditional data processing. This integration marks a significant evolution from extensive computing networks to systems able to think, learn, and act independently, enabling real-time responses and personalised interactions based on historical data and predictive analytics. In doing so, it extends the functionality of ubiquitous computing and further contributes to reducing cognitive load on users by automating complex tasks and providing intuitive interfaces that anticipate user needs. This shift constitutes a leap towards creating environments where technology acts as a proactive assistant in day-to-day activities, fundamentally transforming the way we interact with our digitally augmented surroundings.

2.2 Observing Advanced Practices

2.2.1 Between Internet of Things and Everything

Relevant in this discourse is to distinguish between the two areas of Internet of Things (IoT) and the Internet of Everything (IoE). The IoT refers to the networking of everyday objects, enabling them to send and receive data, essentially making them "smart" [24]. The IoE extends this concept by incorporating three additional components: people, processes, and data [11], hence stressing how connected devices have expanded beyond the traditional framework to include the wider context and environment in which they operate [8, 32, 39]. This concept refers to the system, and integrates these four aspects of daily life to create new capabilities and experiences, significantly enhancing the economic potential for businesses and individuals [23]. The overarching aim is to progressively blend the physical and virtual worlds, leading to a seamless integration of our environments with advanced digital capabilities. This shift emphasises the evolution from merely connected devices to comprehensive ecosystems that leverage interconnected data, processes, and user engagement to drive innovation and efficiency.

For instance, an increasing number of smart light bulbs on the market today allow us to establish routines for turning on and off at specific times or under certain conditions, and even change their colours using a wide chromatic range. These functionalities are managed with extreme intuitiveness and simplicity, controlled via terminals, smartphones, or voice assistants. Technological implementations include ambient light sensors, accelerometers, infrared, and other components that evolve the light bulb into a smart device with decision-making capabilities. The result is a seamless user experience that gathers information, learns, and ultimately reacts to the habits of the people and the usage of the spaces they inhabit. Therefore, key in this reasoning is how current practices are shifting from answering user needs to building user experiences which are not only smooth but immediate, almost with zero access barrier [2, 22].

Observing practices in the home environment, it is a fact that smart systems and devices have become widely diffused [8, 39]. The essence of a smart home lies in the presence of intelligent information and communication technologies, distributed throughout the rooms: interconnected electronic objects that create networked systems capable of transmitting information to users, perceiving their commands, and acting independently to optimise and facilitate home environment management [8, 12]. Energy consumption management represents a particularly crucial area of interest for smart home technologies, becoming the subject of numerous research projects [13, 17, 33, 43]. In this sector, alongside the search for fluid interactions and inter-functionality [32], the issue of high performance is central. Moreover, smart devices must not only be equipped with processors with sophisticated communication capabilities that enable them to communicate with each other and with the external world, but they must also have an infrastructure that is easy to install and configure, requiring little effort for maintenance and management. These aspects have supported the development of minimal interfaces that tend to progressively disappear.

To better understand the challenges associated with the implementation of intelligent systems operating in our home environment, a relevant case is that of the learning thermostats, employing machine learning technology, detection, and connectivity, along with eco-feedback features. With early models introduced in late 2011, they quickly seized the centre of attention from both the media and researchers, exploring how the adoption of adaptive machine learning could allow for the generation of personalised heating and cooling programs based on the habits of the home's users, promoting energy savings.

2.2.2 *Artificial Intelligence of Things*

With technological advances, technology's role transitions from passive to active. Unlike traditional IoT systems that primarily collect, store, and process data, AIoT systems incorporate analysis and control functions, enhancing their functionality. AIoT represents a fusion where AI meets IoT, allowing much of the computational processes to occur at the edge, close to where data is collected by IoT devices.

AIoT systems are characterised by several distinct features [53]:

- Intelligent perception and data collection. IoT devices, through embedded systems and sensors, collect and sense a wide array of environmental data.
- Data processing and analysis. Using AI, the system processes and analyses large volumes of data to extract useful information and identify patterns.

- Autonomous learning and decision making. AI enables IoT devices to autonomously learn from data and make informed decisions, thereby enhancing their adaptability and intelligence.
- Real-time interaction and communication. AIoT devices can interact and communicate in real-time via the Internet, facilitating collaborative processing and task execution among interconnected devices [19].

Within the AIoT ecosystem, the IoT components are critical, being responsible for handling data acquisition and connectivity. They collect environmental data through sensors and facilitate data communication through various protocols like Wi-Fi and Bluetooth. This connectivity is crucial for integrating AIoT into cloud infrastructures where extensive data processing and analysis occur. Consequently, AIoT finds application across numerous sectors, enhancing functionalities in smart homes, health monitoring, smart cities, and intelligent transportation systems. In smart homes, for example, AIoT can personalise experiences by adjusting environmental parameters like lighting and temperature based on user behaviour and preferences, exemplified by devices like AI speakers. These applications demonstrate AIoT's potential to revolutionise everyday life by merging physical and digital functionalities seamlessly, exploiting the potential of smart home ecosystems. Recent models of AI speakers are equipped with advanced voice control capabilities, allowing users to manage home devices such as lighting, door locks, heating systems, and televisions simply through voice commands. Beyond basic control, these AI-powered speakers can act as a hub, integrating with a broader range of smart home devices via Bluetooth Mesh technology. This enables the speaker to not only receive audio content but also to control and coordinate multiple smart devices within the home, enhancing user convenience and efficiency.

As such, AI speakers offer personalised home experiences which are based on learning preferences and behaviours of household members, hence autonomously adjust environmental settings such as temperature, lighting, and even play music based on the learned preferences and habits of specific users. This level of customisation is significant and will further augment, aiming at providing a more intuitive and comfortable living environment, showcasing increased levels of responsiveness to the needs of their inhabitants.

2.2.3 Disappearing Interfaces

The evolving technological landscape illustrates a shift towards smart objects and ecosystems, where interfaces are designed to be increasingly discreet and even hidden, blending into our environments. Following the trend of technology dissolving into the cloud, settings, and objects themselves, interfaces have transformed, often disappearing entirely [22]. This analysis intentionally excludes devices that primarily use voice assistants for interaction, as these do not feature visual interfaces and are beyond the scope of this book.

A prominent example of this trend are interfaces integrated into objects and emerging from surfaces like natural wood, marble, and other materials allowing semi-transparency when engraved. In this context, the concept of calm technology becomes significant [6, 48]. When these devices are inactive, they reveal only a material surface. User interaction activates the display beneath the surface, which then displays commands and information directly on the material itself. The device uses a digital display embedded beneath the surface, appearing at the touch of a hand or through voice command. These also often function as hubs, enabling the sending, receiving, and reading of messages and news, interacting with functions and services, and controlling smart devices (lighting, thermostats, etc.). Once interaction ceases, the display disappears again, reverting to its inanimate appearance. Following the principles of technology that manifests only when needed, these objects do not constantly demand user attention but remain "silent" until their performance is required.

The aim is thus for distraction-free digital communication, allowing users space for normal activities without imposing presence. Interaction occurs both via touch and sound, catering to various needs that shift throughout the day. The focus is not on the technological setup but on the aesthetic aspect. Introducing "non-traditional" interaction modes into daily products and services challenges the underlying assumptions of our common experiences with objects. The resulting increase in products and services with intuitive interfaces leads to a clear shift towards engaging users with different interaction ecologies, activating new social and cultural rituals. This necessitates rethinking how we build and use technological artefacts, considering benefits from the user's perspective, as well as the desire for innovation that drives designers. Therefore, the real question unfolds in a variety of equally fundamental aspects. What are the consequences of introducing interactions based on grammars other than those we have learned to use and manage so far? To what extent are we willing to give up our data and thus our privacy in exchange for greater comfort?

2.3 Reflecting on Gaps and Challenges

The emerging landscape is increasingly connected and interconnected, offering designers, engineers, and entrepreneurs new opportunities for innovation and development. However, alongside the array of intriguing possibilities, it is a duty to continually pose questions and especially reflect on the implications that new technologies bring: economically, socially, and culturally. The analysis of the state of the art so far presented provided a glimpse of some advancements in this field, pointing out current challenges related to how interfaces are perceived, and gaps both in academic and practical terms nurtured by the constant technological advances—see the implication of AIoT.

Building upon this discourse requires necessarily opening a reflection on the varying degrees of visibility associated with interfaces, considering the spectrum that ranges from highly visible and recognisable interfaces to those that are nearly invisible in their operation [29]. This exploration includes pointing out an initial reasoning about the implication behind interfaces designed to be almost invisible within the user's environment. It opens relevant issues for discussing how such interfaces can blend seamlessly into the background of everyday life, altering the conventional dynamics between users and technology. By becoming almost invisible, these interfaces shift the focus from direct interaction to more ambient and intuitive forms of user engagement [6, 20]. The discourse explores this transition, considering how it affects usability, accessibility, and the overall relationship between humans and technology. This analysis offers the fundamentals for the next chapters.

Moreover, the integration of smart objects and systems and their underpinning technologies into our daily environments comes with challenges and gaps, which require a holistic and multidisciplinary approach. First, interoperability and standardisation remain significant hurdles as devices from different manufacturers often fail to communicate seamlessly, undermining the potential of smart ecosystems [42]. Privacy and security are also critical issues, with concerns related to data breaches and user trust towards smart and networked systems [38]. Reflecting on the perceptions and management of privacy risks in smart homes, several key themes emerge from the literature [54]. For instance, the preference for convenience often shapes user behaviour towards privacy, particularly in their interactions with external parties: they tend to both trust how IoT devices will use their data, and prioritise benefits over potential privacy risks, impacting their attitudes towards data collection.

Furthermore, governance issues present another complex challenge, as the development and implementation of consistent policies and regulations across different regions and sectors are crucial to ensure the security, privacy, and ethical use of smart technologies [44]. While recognising how ensuring intuitiveness and user-friendliness is crucial, it is also necessary to consider the risk of alienation among certain users who may not be tech-savvy, including the elderly, or have impairments which prevent specific typology of interaction. The economic and social disparities that smart technologies might exacerbate also pose a challenge, highlighting the need for strategies to bridge the digital divide and ensure equitable access [21]. Last but not least, the ethical implications of reduced human oversight in decision-making processes remains a key issue, with autonomy shifts to AI systems which call for robust and shared ethical guidelines and governance [5, 26]. These issues represent current challenges and gaps identified in both the literature and practical fields, requiring further investigation. Addressing these issues requires not only technological solutions but also insights from design, ethics, and social sciences to create systems that are secure, fair, and beneficial for all users.

References

1. M.K. Afzal, Y.B. Zikria, S. Mumtaz, A. Rayes, A. Al-Dulaimi, M. Guizani, Unlocking 5G spectrum potential for intelligent IoT: opportunities, challenges, and solutions. IEEE Commun. Mag. **56**(10), 92–93 (2018). https://doi.org/10.1109/MCOM.2018.8493125
2. K. Ashton, That "Internet of Things" thing. RFID J. (2009). https://www.rfidjournal.com/articles/view?4986
3. N. Balta-Ozkan, R. Davidson, M. Bicket, L. Whitmarsh, Social barriers to the adoption of smart homes. Energy Policy **63**, 363–374 (2013). https://doi.org/10.1016/j.enpol.2013.08.043
4. P. Bansal, E. Vineyard, O. Abdelaziz, Advances in household appliances- a review. Appl. Therm. Eng. **31**(17), 3748–3760 (2011). https://doi.org/10.1016/j.applthermaleng.2011.07.023
5. S.E. Bibri, Ethical implications of AmI and the IoT: risks to privacy, security, and trust, and prospective technological safeguards, in *The Shaping of Ambient Intelligence and the Internet of Things: Historico-epistemic, Socio-cultural, Politico-institutional and Eco-environmental Dimensions*, ed. by S. E. Bibri (Atlantis Press, Paris, 2015), pp. 217–238. https://doi.org/10.2991/978-94-6239-142-0_7
6. A. Case, *Calm Technology: Principles and Patterns for Non-intrusive Design* (O'Reilly Media, Sebastopol, CA, 2015)
7. M. Castells, *The Network Society: A Cross-Cultural Perspective* (Edward Elgar, London, 2004)
8. D.J. Cook, How smart is your home? Science **335**(6076), 1579–1581 (2012). https://doi.org/10.1126/science.1217640
9. Y. Dong, Research on logistics talents cultivation strategy against the background of artificial intelligence and internet of things, in *Proceedings of the 2020 Conference on Education, Language and Inter-cultural Communication (ELIC 2020)* (Atlantis Press, 2020), pp. 1–6. https://doi.org/10.2991/assehr.k.201127.001
10. D. Evans, The internet of things: how the next evolution of the internet is changing everything, in *Cisco Internet Business Solutions Group (IBSG)* (2011), pp. 1–11 . www.cisco.com/c/dam/en_us/about/ac79/docs/innov/IoT_IBSG_0411FINAL.pdf
11. D. Evans, The internet of everything: how more relevant and valuable connections will change the world, in *Cisco Internet Business Solutions Group (IBSG), 2012* (2012), pp. 1–9. https://www.cisco.com/c/dam/global/en_my/assets/ciscoinnovate/pdfs/IoE.pdf
12. S.K. Firth, F. Fouchal, T. Kane, V. Dimitriou, T.M. Hassan, Decision support systems for domestic retrofit provision using smart home data streams, in *Proceedings of the 30th International Conference on Applications of IT in the AEC Industry. Move Towards Smart Buildings: Infrastructures and Cities, Beijing, China* (2013)
13. J. Froehlich, L. Findlater, J. Landay, The design of eco-feedback technology, in *Proceedings of the SIGCHI Conference on Human Factors in Computing Systems* (2010), pp. 1999–2008
14. L. Gonçalves, L. Patrício, J. Grenha Teixeira, N.V. Wünderlich, Understanding the customer experience with smart services. J. Serv. Manag. **31**(4), 723–744 (2020). https://doi.org/10.1108/JOSM-11-2019-0349
15. C. Goumopoulos, I. Mavrommati, A framework for pervasive computing applications based on smart objects and end user development. J. Syst. Softw. **162**, 110496 (2020). https://doi.org/10.1016/j.jss.2019.110496
16. F. Gullì S. Ceccacci, M. Germani, L. Cavalieri, Design adaptable and adaptive user interfaces: a method to manage the information, in *Ambient Assisted Living: Italian Forum 2014*, ed. by B. Andò, P. Siciliano, V. Marletta, A. Monteriù (Springer International Publishing, Cham, 2015), pp. 47–58. https://doi.org/10.1007/978-3-319-18374-9_5
17. M. Gupta, SS. Intille, K. Larson, Adding gps-control to traditional thermostats: an exploration of potential energy savings and design challenges, in *International Conference on Pervasive Computing* (Springer, 2009), pp. 95–114
18. K. Gushima, Y. Kinoshita, T. Nakajima, Pervasive smart objects: framework for extending smart-object services, in *Distributed, Ambient and Pervasive Interactions*. ed. by N. Streitz, S. Konomi (Springer International Publishing, Cham, 2021), pp.100–121

19. A. Haroun, X. Le, S. Gao, B. Dong, T. He, Z. Zhang et al., Progress in micro/nano sensors and nanoenergy for future AIoT-based smart home applications. Nano Express **2**(2), 022005 (2021). https://doi.org/10.1088/2632-959X/abf3d4
20. M. Hassenzahl, N. Tractinsky, User experience—a research agenda. Behav. Inf. Technol. **25**(2), 91–97 (2006). https://doi.org/10.1080/01449290500330331
21. B. Hofmann, Ethical challenges with welfare technology: a review of the literature. Sci. Eng. Ethics **19**(2), 389–406 (2013). https://doi.org/10.1007/s11948-011-9348-1
22. T.K. Hui, R.S. Sherratt, Towards disappearing user interfaces for ubiquitous computing: human enhancement from sixth sense to super senses. J. Ambient. Intell. Humaniz. Comput. **8**(3), 449–465 (2017). https://doi.org/10.1007/s12652-016-0409-9
23. F. Hussain, Internet of everything, in *Internet of Things: Building Blocks and Business Models*, ed. by F. Hussain (Springer International Publishing, Cham, 2017), pp. 1–11. https://doi.org/10.1007/978-3-319-55405-1_1
24. T. Jenkins, I. Bogost, Designing for the internet of things: prototyping material interactions, in *CHI'14 Extended Abstracts on Human Factors in Computing Systems* (ACM, 2014), pp. 731–740. https://doi.org/10.1145/2559206.2578879
25. H. Kang, K.J. Kim, Feeling connected to smart objects? A moderated mediation model of locus of agency, anthropomorphism, and sense of connectedness. Int. J. Hum Comput Stud. **133**, 45–55 (2020). https://doi.org/10.1016/j.ijhcs.2019.09.002
26. A. Karale, The challenges of IoT addressing security, ethics, privacy, and laws. Internet Things **15**, 100420 (2021). https://doi.org/10.1016/j.iot.2021.100420
27. M.J. Kim, M.E. Cho, H.J. Jun, Developing design solutions for smart homes through user-centered scenarios. Front. Psychol. **11**, 335 (2020). https://doi.org/10.3389/fpsyg.2020.00335
28. M. Kranz, P. Holleis, A. Schmidt, Embedded interaction: interacting with the internet of things. IEEE Internet Comput. **2**, 46–53 (2009). https://doi.org/10.1109/MIC.2009.141
29. G. Krishna, *The Best Interface is No Iinterface: The Simple Path to Brilliant Technology* (New Riders, San Francisco, CA, 2015)
30. M. Kuniavsky, *Smart Things, Ubiquitous Computing User Experience Design* (Elsevier, Burlington, MA, 2010)
31. B. Latour, On interobjectivity. Mind Cult. Act. **3**(4), 228–245 (1996)
32. Y.-J. Lin, H.A. Latchman, M. Lee, S. Katar, A power line communication network infrastructure for the smart home. IEEE Wirel. Commun. **9**(6), 104–111 (2002)
33. I. Machorro-Cano, G. Alor-Hernández, M.A. Paredes-Valverde, L. Rodríguez-Mazahua, J.L. Sánchez-Cervantes, J.O. Olmedo-Aguirre, HEMS-IoT: a big data and machine learning-based smart home system for energy saving. *Energies 13*(5). https://doi.org/10.3390/en13051097
34. D. Marikyan, S. Papagiannidis, E. Alamanos, Cognitive dissonance in technology adoption: a study of smart home users. Inf. Syst. Front. **25**(3), 1101–1123 (2023). https://doi.org/10.1007/s10796-020-10042-3
35. M. McLuhan, *Understanding Media: The Extensions of Man* (The New American Library, New York, 1964)
36. M.H. Miraz, M. Ali, P.S. Excell, R. Picking, A review on Internet of Things (IoT), Internet of everything (IoE) and Internet of Nano Things (IoNT), in *2015 Internet Technologies and Applications (ITA)* (IEEE, 2015), pp. 219–224
37. A.A. Nazari Shirehjini, A. Semsar, Human interaction with IoT-based smart environments. Multimed. Tools Appl. **76**(11), 13343–13365 (2017). https://doi.org/10.1007/s11042-016-3697-3
38. M.M. Ogonji, G. Okeyo, J.M. Wafula, A survey on privacy and security of internet of things. Comput. Sci. Rev. **38**, 100312 (2020). https://doi.org/10.1016/j.cosrev.2020.100312
39. S.H. Park, S.H. Won, J.B. Lee, S.W. Kim, Smart home–digitally engineered domestic life. Pers. Ubiquitous Comput. **7**(3–4), 189–196 (2003). https://doi.org/10.1007/s00779-003-0228-9
40. Ö. Raudanjoki, J. Häkkilä, M. Pakanen, A. Colley, Shadow display design concepts for AI enhanced environments, in *ArtsIT, Interactivity and Game Creation*, ed. by A.L. Brooks (Springer Nature Switzerland, Cham, 2023), pp. 374–388

41. I. Røpke, T.H. Christensen, J.O. Jensen, Information and communication technologies–A new round of household electrification. Energy Policy **38**(4), 1764–1773 (2010). https://doi.org/10.1016/j.enpol.2009.11.052
42. J. Saleem, M. Hammoudeh, U. Raza, B. Adebisi, R. Ande, IoT standardisation: challenges, perspectives and solution, in *Proceedings of the 2nd International Conference on Future Networks and Distributed Systems.* (Association for Computing Machinery, New York, NY, USA, 2018). https://doi.org/10.1145/3231053.3231103
43. J. Scott, A. Bernheim Brush, J. Krumm, B. Meyers, M. Hazas, S. Hodges, N. Villar, PreHeat: controlling home heating using occupancy prediction, in *Proceedings of the 13th International Conference on Ubiquitous Computing* (2011), pp. 281–290
44. A. Sebastian, S. Sivagurunathan, V. Muthu Ganeshan, IoT challenges in data and citizen-centric smart city governance, in *Smart Cities: Development and Governance Frameworks* ed. by Z. Mahmood (Springer International Publishing, Cham, 2018), pp. 127–151. https://doi.org/10.1007/978-3-319-76669-0_6
45. J. Sweller, Cognitive load theory, learning difficulty, and instructional design. Learn. Instr. **4**(4), 295–312 (1994)
46. J. Tidwell, *Designing Interfaces: Patterns for Effective Interaction Design* (O'Reilly Media, Sebastopol, CA, 2010)
47. O. Vermesan, P. Friess, *Internet of Things: Converging Technologies for Smart Environments and Integrated Ecosystems* (River publishers, Delft, The Netherlands, 2013)
48. M. Weiser, J.S. Brown, Designing calm technology. PowerGrid J. **1**(1), 75–85 (1996)
49. M. Wiberg, H. Ishii, P. Dourish, D. Rosner, A. Vallgårda, P. Sundström, et al., Material interactions: from atoms & bits to entangled practices, in *CHI'12 Extended Abstracts on Human Factors in Computing Systems* (ACM, 2012), pp. 1147–1150
50. M. Wiberg, E. Robles, Computational compositions: aesthetics, materials, and interaction design. Int. J. Des. **4**(2), 65–76 (2010)
51. C. Wilson, T. Hargreaves, R. Hauxwell-Baldwin, Smart homes and their users: a systematic analysis and key challenges. Pers. Ubiquit. Comput. **19**(2), 463–476 (2015). https://doi.org/10.1007/s00779-014-0813-0
52. E. R. B. de Wolf, Pervasive technology as scaffolds for mindful living: reframing calm technology (2021). http://essay.utwente.nl/88238/
53. J. Yan, AIoT in smart homes: challenges, strategic solutions, and future directions. Highlights Sci. Eng. Technol. **87**, 59–65 (2024). https://doi.org/10.54097/8hzgaf51
54. S. Zheng, N. Apthorpe, M. Chetty, N. Feamster, User perceptions of smart home IoT privacy. Proc. ACM Hum.-Comput. Interact **2**(CSCW). https://doi.org/10.1145/3274469

55. [illegible]

Chapter 3
Technology Transfer. Bringing Scientific Research to the Market

Abstract This chapter explores Think's pathway for translating theoretical research into research prototype and market-ready applications within the context of smart objects and IoT products. It discusses the pivotal role of incubators like PoliHub in nurturing technology startups and delves into the process of bridging the gap between academic research and commercial viability. The discourse reports on Thingk's unique approach to product design, adopting an interdisciplinary and integrated perspective.

Keywords Technology transfer · Spin-off · Incubation · Market application · Research and innovation

3.1 Technology Transfer

The chapter explores the technology transfer of Thingk, a spin-off from Politecnico di Milano that experiments with interactive technologies, designing products that disguise themselves. Thingk's foundational approach is to craft artefacts that, while appearing simple and minimal in form, are imbued with advanced technological functions, following the slogan: objects of daily use with superpowers.

This chapter focuses on the technology transfer and product innovation, critically analysing Thingk's technology transfer in the field of smart objects and IoT products that originated from fundamental research in the domains of interaction design, product design, and communication design. These fields, in a synergistic interplay with computer science and engineering, have nurtured the conception of novel concepts of embodied interfaces playing with shape and function. The analysis of the research landscape and market led to identifying research outcomes that have potential practical applications. Rather than pinpointing technologies that can address real-world problems or needs and having a vertical view on current needs, the idea was to follow the disruptive and challenging concept of **augmenting objects with superpowers.**

U. Tolino and I. Mariani, *Design Behind Interaction*,
PoliMI SpringerBriefs, https://doi.org/10.1007/978-3-031-67416-7_3

Technology transfer is the process through which knowledge, technologies, production methods, prototypes, and services developed by governments, universities, corporations, and both public and private research institutions can be made accessible to a wide range of users who can then further develop and exploit the technology to create new products, processes, applications, materials, and services [21]. It is thus an open innovation methodology [6, 17, 29], comprising a set of activities aimed at transferring knowledge (technology, skills, manufacturing methods, production samples, and services) from the world of scientific research to the market, making technology accessible to ecosystems external to the universities where they are developed.

The process of creating an idea and the transfer phase involve an ecosystem of stakeholders: the institutions and actors (researchers) who are engaged in research and development activities, and therefore the research and innovation carried out in universities; small and medium enterprises and companies that participate to varying extents in research projects—in our case, especially highly specialised manufacturers and technology providers—that activate paths of open innovation outside the corporate context; public and private funding bodies that support with external financial resources; and the Technology Transfer Offices (TTOs) as specialised actors in transfer that act as a bridge between the world of research and the market [3, 22], supporting researchers' efforts to position research programs in the market, commercialise intellectual property, and provide information on market trends.

Drawing upon a theoretical **framework** that intersects various dimensions of the technology transfer process [21], the case of Thingk offers critical insights into the dynamic interplay between academic research and market application in the context of smart technology (Table 3.1).

Building upon the relevant dimensions outlined in Table 3.1, which have been derived from the scientific discourse on the subject, what follows explores how each dimension of the theoretical framework is operationalised within Thingk's activities, furthering innovation. As such, Thingk becomes a lens to explore the multifaceted and often non-linear nature of technology transfer.

3.1.1 Contribution of Science Parks and Incubator

In discussing the support mechanisms for technology-based startups and spin-offs, critical is the role of research, science, and technology parks as key vehicles for technology transfer [24]. These parks, including PoliHub—an innovation park and startup accelerator associated with the Politecnico di Milano—serve not only as physical hubs that foster innovation but also as facilitative environments for turning scientific research into market-ready technologies [13]. In promoting development, parks and incubators are adopting different approaches ranging from "laissez-faire" to "strong intervention" [5], tailoring their support to meet the diverse needs of startups at different stages of their maturity. Moreover, beyond supporting the development of start-ups and spin-offs, these parks increasingly focus on **promoting local**

Table 3.1 Theoretical framework related to Thingk's operations within the technology transfer (TT) context

	Dimension	References	Role in thingk
D01	Contribution of science parks and incubator	Support for technology-based firms: [5, 7, 11, 13, 24] Local economic development: [8, 23]	Thingk was born as a spin-off within the incubator of the Politecnico di Milano (PoliHub), that provided resources, networking opportunities, knowledge programmes, and business support services
D02	Venture funds	R&D fundings: [4, 31] (Coupet 2021)	Thingk has largely participated in various national and international fundings through competitive bids for speeding up TT activities, securing relevant economic support for its research and experimentation
D03	People and organisational culture	Human capital and culture: [16, 37]	Thingk's team comprises nine individuals, including four founders, from diverse disciplinary backgrounds with complementary skills and expertise. The environment is open to continuous dialogue and experimentation that welcomes both researchers and practitioners

(continued)

Table 3.1 (continued)

	Dimension	References	Role in thingk
D04	TT ecosystem structure	Organisational structure: [3, 33, 34] Activities by TTOs, science parks, incubators, and venture funds: [2, 25, 32] (Neves & Franco 2016) Holistic consideration of the TT ecosystem: [18]	Embedded within the PoliHub incubator, Thingk leveraged its fundamental knowledge and experience to establish a clear strategic direction and assets to support growth. The incubator provides complementary skills, assets, and expertise otherwise not included in the spin-off team. Additionally, Thingk is rooted in crowdfunding, co-creation and co-design with end-users and relevant stakeholders, building a small ecosystem with a dynamic governance among the parties Thingk actively participated in various initiatives at the incubator PoliHub. This includes participation in technology commercialisation, networking, securing national and international funding, and determining commercialisation strategies The incubator environment provided Thingk with continuous interaction and synergy with other spin-offs and key industry actors. The incubator setting as an ecosystem supported continuous synergies and exchanges with other spin-offs as well as relevant actors in the field of Design, Smart products and home appliances communities
D05	TTO role	TTOs as a bridge: [19, 25, 33, 37] Protecting university rights and supporting pre-commercialization: [36]	Thingk worked closely with the TTO to bridge the gap between academia and the market. The TTO provided essential support for the preparatory documentation and activities to the patent, particularly in terms of prior art search and intellectual property strategy

economic development, thus contributing to bringing change and innovation into a broader ecosystem. In this sense, **university-based technology transfer is recognised to** play a pivotal role in regional economic development by facilitating the flow of innovations from academia to the market and its actors. Significant is the contribution of **PoliHub** in providing entrepreneurial support [1]. PoliHub offered not only physical resources such as workspace and technological infrastructure but also crucial networking opportunities that connected Thingk with potential collaborators, investors, and industry experts. The incubator also offered various knowledge programs, and specifically mentorship and training that equipped the Thingk team with complementary knowledge in areas like business management, intellectual property rights (via TTO), and commercial strategy.

3.1.2 Fundings and Academic Entrepreneurship

Building on this foundational role, universities have increasingly embraced entrepreneurial roles, especially through the expansion and intensification of technology transfer activities, which involve commercialising technologies developed within university settings. This shift has been driven partly by supportive policy environments that favour patenting and licensing, alongside the need to identify a position and hence steer the competitive landscape for **funding resources** [31]. Supported by different strands of funding, technological universities act as critical hubs for knowledge and technology dissemination, fostering local industry advancements, often enhancing local and regional economic growth by favouring collaborations with actors from the surrounding ecosystem for translating academic research into commercially viable innovation, shaped as products, services, and systems [8, 23]. Operating within PoliHub, Thingk has leveraged entrepreneurial and technological resources made available by the academic context, favouring the advancement of both research and commercialisation efforts. The support from PoliHub has been instrumental in facilitating Thingk's access to national and international funding, which has accelerated its technology transfer activities. This strategic use of resources underlines the critical role of university-incubated enterprises in translating theoretical research into market-ready solutions, highlighting the tangible benefits of aligning academic pursuits with commercial opportunities and societal needs.

3.1.3 People and Organisational Culture

To exploit the technology transfer while better tackling the challenges emerging from the smart home appliances domain, the spin-off was born with an interesting mix of competences which take advantage of the diversity in expertise and backgrounds of its members. Exactly the composition and culture of Thingk's team has fueled the spin-off's innovative capacity [33, 34] and responsiveness and even anticipation of market needs. The synthesis of complementary skills and knowledge within a multidisciplinary team has created a dynamic environment conducive to dialogue and experimentation, that welcomes both researchers and practitioners and plays a crucial role in fostering innovative capacities and responsiveness to market dynamics [16, 37]. The Think team is composed of nine individuals, all alumni from Politecnico di Milano spanning disciplines such as design (interaction, product, communication), electronic and computer science engineering. The multidisciplinary nature of the team plays a crucial role in transitioning from research to market-ready innovations. Specifically, the diversity of the academic background and professional experiences in design and technology not only enhances the ability to address complex problems but also enriches the process of knowledge and technology transfer. This integration of multiple perspectives ensures that the innovations developed are not only comprehensive but also deeply integrated with practical applications, reflecting a balanced understanding of technical feasibility and user-centric design. Moreover, each team member brought additional expertise to the spin-off, encompassing fields such as management, sociology, anthropology, ethnography, and marketing. This diversity broadened the team's perspective, enhancing their ability to understand and predict market trends and user behaviours. It also enabled them to independently manage the entire design process—from concept development through data analysis to user validation—ensuring a comprehensive approach to product development and market entry. This diverse expertise facilitates a more holistic view of product development, from initial design to market entry, ensuring that each innovation is not only technically robust but also culturally relevant.

Thingk's organisational culture is horizontal, fostering an inclusive environment where every viewpoint is valued and encouraged in order to ensure that a wide array of perspectives can inform the development process, enhancing creativity and innovation. Furthermore, while there is a general design process that the spin-off adheres to—as detailed in Chap. 4—this process is flexible and can be customised depending on the specific requirements of the innovation at hand.

3.1.4 An Extended and Comprehensive Technology Transfer Ecosystem

Research into academic entrepreneurship and technology transfer underscores indeed the importance of strategic interactions within the technology transfer ecosystem [18, 33]. It highlights the relevance of such an ecosystem, with actors offering strategic advantages that enhance productivity and foster organisational learning, aligning university research with industry partners and policy stakeholders [14, 32]. The discourse also highlights how economic, social, and political factors influence the universities' ability to innovate and apply knowledge in ways that foster economic growth and offer broader societal benefits [4, 27].

The alignment of a **spin-off's goals with the components of the technology transfer ecosystem** can significantly influence its success and trajectory.

The integration of academic goals with market needs [25], understanding regulatory frameworks, and facilitating equitable technology being aware of the user's different digital literacies in the emerging setting of IoT and smart objects are essential challenges in technology transfer.

Thingk benefits significantly from a supportive environment provided by PoliHub, which offered resources crucial for tackling these challenges.

Favourable conditions come from the overall ecosystem, ranging from highly supportive institutions and their members to small and medium enterprises and companies, including highly specialised manufacturers and technology providers which engaged in open innovation paths outside traditional corporate frameworks and actively participated in experiments with the spin-off. Additionally, the spin-off features a consistent and active engagement with various stakeholders, including users who contributed insights and co-designed solutions throughout the development process. Thingk's journey is also remarkable for its innovative approach to **governance and stakeholder** engagement [26]. The adoption of crowdfunding and co-creation strategies, facilitated user engagement and iterative product development [20], albeit with inherent challenges. Evident are however the benefits behind reaching out to a worldwide audience in a prototypical phase, who engaged about 400 users for testing and co-creation, contributing with bottom-up value creation [10, 12].

These interactions are supported by both public and private funding bodies that provide essential financial resources, reinforcing the spin-off's activities. The spin-off underwent a the shift from initial competitive national grants for supporting experimentation shaped as a first production of smart objects to the European competitive fundings, synergies, and partnership with relevant actors [15, 18] for advancing the experimentation and exploring emerging technologies [2, 9, 25, 28, 32]. This support boosted Thingk's growth, weighing their contribution against challenges and limitations inherent in such entrepreneurial ecosystems [5, 7, 11, 13, 27]. Moreover, the TTOs served as a constant guide towards the complexities of commercialisation and patenting [2, 3, 34].

3.1.5 The Role of the TTO

The TTO has been pivotal in facilitating the transition from academic research to market-ready innovations, serving as crucial bridges between academia and industry. Thingk's largely benefitted from TTO support in dealing with intellectual property rights and commercialisation strategies. Notably, the TTO's involvement was instrumental in preparing the necessary documentation and conducting prior art searches, thereby ensuring that Thingk's innovations were adequately protected and strategically positioned within the market. This relationship underscores the TTO's broader function of protecting university rights while actively supporting the pre-commercialisation phases of development [19, 25, 33, 36, 37]. Such strategic partnerships help bridge the gap between innovative research and its practical application and operationalisation, enhancing the potential for economic impact through technology transfer.

3.2 Product Development

Building upon the discourse presented thus far, the discussion below explores the process, challenges, and opportunities arising from the complex dynamics involved in transitioning from theoretical research to market readiness. It highlights the development of seven products featuring hidden interfaces, with three currently on the market and four as research prototypes. This exploration includes the innovative concept of "drawing interfaces". In doing so, it is explored the critical role that different entities played in this transformative phase [5, 30, 35, 38].

Building upon the technology transfer principles outlined earlier, Thingk reimagines everyday objects by integrating interactive technologies, a methodology and process detailed in Chap. 4. To contextualise Thingk's innovation within the broader dialogue on IoT for smart home appliances, it is essential to revisit the foundational principles of Thingk's innovation and the roots of its development. This backdrop sets the stage for understanding how Thingk's innovations align with and contribute to the evolving landscape of smart technology. By merging archetypal geometric shapes with digital capabilities, Thingk pursues a sense of innovation that is based on hacking the meaning of seemingly analogical, basic, and minimal products, turning them into everyday objects endowed with superpowers. All products are crafted by studying the processing of natural materials (wood, leather, marble, stone) in collaboration with the suppliers and manufacturers themselves, to achieve an analog look and feel that conceals a technological soul (the superpowers) ready to emerge when needed. The objects' aesthetics (shape) do not convey their function (affordance). Thingk aims for innovation that surprises by playing with the form-function relationship, combining formal simplification and enhanced functionality. By inserting interfaces that remain silent until interacted with, Thingk empowers and augments objects to declare their meaning: the relationship between form and function ceases to be predictable or familiar, relying instead on a language that is anything but taken for granted.

This philosophy led us to design three objects for our target market: an intelligent scale that appears to be a birch cutting board (*Slab!*, 2014), a wireless smartphone charger that resembles a beechwood disc (*Disc!*, 2015), and a wireless charging pad integrated into a leather desk mat (*Desk!*, 2015).

Fig. 3.1 *Slab!* scale/timer, and *Disc!* and *Desk!* wireless desk chargers

Simultaneously, during research activities within the H2020 European project DecoChrom,[1] which focuses on developing innovative, chromogenic technology applications, experimentation advanced the technological evolution of *Slab!* (*Slab! XL*) e *Disc!* (*Disc! XL*), leading to the conception of two innovative, patent-pending products: a simplified wireless hard drive for transferring single files (*Coin!*) and a controller with a variable interface for managing home automation functions (*Knob!*).

In each of these products, it is evident that the design intentionally conceals their true functionality, playing on the user's initial perceptions and assumptions, leading to a deliberate reveal and user interaction.

3.2.1 Market Production

Thingk's production has led to the creation of a line of products for the home and office, characterised by minimalist designs that appear to have a single function (suggested by their form) but conceal at least one secondary smart function enabled by the adopted technology (Fig. 3.1).

Slab! is a kitchen scale, a digital timer, and an IoT device with Bluetooth connectivity for future integrations with other smart devices. It hides its functional complexity with a minimalistic and natural aesthetic. It has been designed with the awareness that the choice of natural material enriched with hidden electronic

[1] DecoChrom—Decorative Applications for Self-Organized Molecular Electrochromic Systems, was a EU research project funded under the H2020 program, call H2020-EU.2.1.3.—INDUSTRIAL LEADERSHIP—Leadership in enabling and industrial technologies—Advanced materials. GA. 760973, duration 2018-2022. Link: cordis.europa.eu/project/id/760973; decochrom.com.

components can surprise users during various interactions. In a horizontal position, it functions as a scale; in a vertical position, it transforms into a timer. When placed on its sides, it allows the activation of secondary functions, such as Bluetooth or the change of measurement scale. The information is consistently provided using the single digital display located beneath a layer of wood milled to less than one millimetre in thickness.

As a result, this smart kitchen scale is conceived to create a user experience that blends the physical and digital realms. Its appearance is essential and natural. The high-quality wood gives it an elegant and minimalist profile. Its seemingly analog appearance conceals a digital nature and functional complexity that literally emerge from the surface of the object.

Disc! is a wireless charging base made entirely of natural wood milled to very thin thicknesses, allowing the transfer of electric charge upon contact with a smartphone. In its advanced version, ***Disc! XL***, designed as a research prototype (description in Sect. 3.2.2), doubles in size, adding the functionality of a smart power bank that allows the transport of an additional battery without the need for an electrical outlet.

Desk! is an elegant desk mat crafted from leather by high-end artisans from the productive district of Brianza. Beneath its cover, it conceals two wireless charging zones that react only to contact with a smartphone placed on top of them, utilising standard QI charging technology. This object maintains the original desk mat style, enhancing its capabilities only when needed, through embossed graphics that reveal the areas where the desk mat performs its function.

Within the broader discourse of this chapter, the development of products such as *Slab!*, *Disc!*, and *Desk!* exemplifies the practical application of technology transfer principles. These products are conceived by transforming research outcomes into tangible consumer goods, demonstrating technology transfer as a catalyst for innovation. This process enables academic research to transition into commercially viable solutions that effectively address real-world needs.

3.2.2 Research Prototypes

As previously mentioned, ***Slab!*** and ***Disc!*** were selected as starting points for experimentation within the H2020 European project DecoChrom. The project conducts thorough experimentation with electrochromic inks to create functional displays on everyday objects, aligning seamlessly with the design philosophy behind Thingk's products. This context offers a fertile and privileged setting for enhancing the original performance of *Slab!* and *Disc!* by utilising new sensors and the mimetic properties allowed by electrochromic technology (Fig. 3.2).

For the ***Slab! XL*** prototype, the primary objective is to improve the user experience, particularly addressing the main cognitive challenges encountered in the interaction with the original Slab!. Users frequently encountered difficulties in understanding the shifts between its multiple functions—such as transitioning from a scale to a timer, or engaging Bluetooth connectivity—details of which are elaborated in

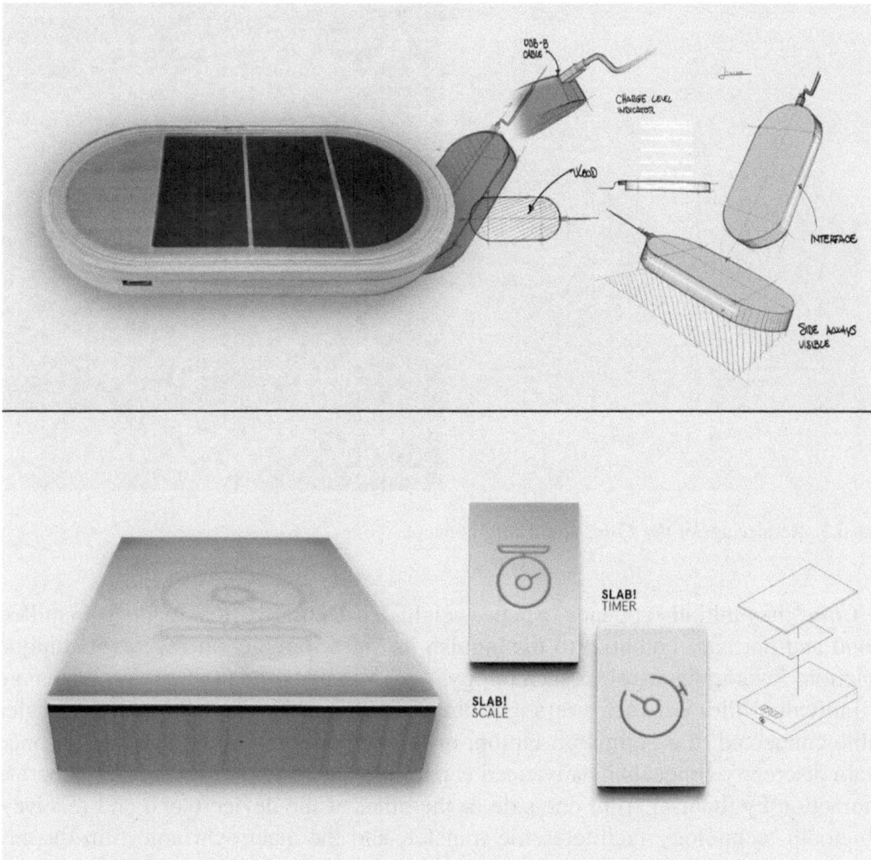

Fig. 3.2 *Slab! XL* and *Disc! XL* prototypes with electrochromic display and renderings of the *Coin!* and *Knob!* concepts

Chap. 4. To differently tackle these challenges, we tested the introduction of a thin and transparent electrochromic display that was adopted to showcase iconic symbols directly on the wooden surface, clearly indicating the corresponding functions based on the product's orientation and use.

For the ***Disc! XL*** prototype, the focus is on enhancing the interaction between the wooden surface and another device using a sensor responsive to ambient noise, interpreting it as an interactive trigger. This approach addresses the user-reported need for confirmation feedback upon the start of charging, displayed through graphics that appear on the object's surface, resembling a battery icon.

Continuing within the DecoChrom research framework, two additional prototypes were specifically designed to highlight the unique characteristics of electrochromic displays (Fig. 3.3).

Fig. 3.3 Renderings of the *Coin!* and *Knob!* concepts

Coin! may initially appear as a paperweight or a decorative item made from milled wood and partially coloured to distinguish its sides. Interaction reveals its unique function: leveraging wireless technology, *Coin!* enables the storage and exchange of individual files between users and devices without the need for a data transfer cable connected to a computer, laptop, or smartphone. Its minimalist design once again discreetly conceals its advanced capabilities, activated only after a deliberate interaction by the user from one side or the other of the device (send and receive). Bluetooth technology facilitates file transfer, and the electrochromic film display indicates the transfer's completion or the presence/absence of files within it.

Knob! is a cylindrical controller incorporating proximity sensors that allow the adjustment of various values of parameters of surrounding devices, such as temperature, light intensity, and volume. The interface of the object comprises multiple layers, each with its own function. However, it remains inactive in its idle state and does not display its functions until placed in a specific context, subsequently showing the commands relevant to that scenario. This behaviour feature endows the interface with performative capabilities, where its visual elements can dynamically act and react to user manipulation, rendering the displayed information and functions "variable" based on the usage context.

3.3 Patent Filing

Following the experimentation within the DecoChrom project, the research prototype *Coin!* underwent further development and refinement, which led to the filing of a European Design Patent.[2]

The European Design Patent represents and epitomises the direct exploitation of the research and experimentation undertaken, particularly leveraging the extensive prototyping, user engagement, and analysis performed throughout the piloting activities. This patent stands as a testament to the strategic and practical applications of academic research, reflecting its exploitation potential in terms of technological advancements. It highlights how the insights gained from user feedback and iterative design processes have been harnessed to create a product that merges academic rigour with commercial viability, thereby contributing to the advancement of smart technology in consumer products.

3.4 Innovation Seeds

The spin-off Thingk has effectively leveraged the founders' ability to convert innovative ideas into market-ready solutions, often predicting and addressing emerging market needs. This process includes the exploration of promising research directions, identifying gaps in the market, and consequently conceptualising potential products. This means maintaining adherence to the scientific approach, while aligning with market demands, and ensuring technological feasibility. Moreover, the integration of smart technologies should also require a thorough evaluation of ethical considerations, ensuring that these advancements are used responsibly and benefit society as a whole.

As presented so far, a cornerstone of Thingk's entrepreneurial strategy is the emphasis on applied experimentation, characterised by an inherently and highly iterative process grounded in theoretical principles and cutting-edge technologies. In so doing, Thingk systematically engages with theoretical constructs and leverages technologies to nurture initial concepts and turn them into fully-fledged products. Thingk addresses the gap between academic research and market application, adopting a strategic and systematic approach, marked by a continuous commitment to co-design and active engagement with key users, stakeholders and innovators within the smart home appliances sector. This inclusive strategy spans interactions with early adopters and prosumers, highly specialised manufacturers and technology providers, ensuring a comprehensive integration of diverse expertise and insights.

---

[2] European Design Patent: *Coin!* Portable wireless archive, No. 102021000027413—Application: 26/10/2021; Publication: 26/04/2023. Inventors: Stefano Marangoni, Umberto Tolino, Ilaria Mariani. Concept and product design of an interactive wireless solid-state portable memory and a data transfer method using such portable memory.

Link: patents.google.com/patent/IT202100027413A1/en.

The result of this approach has led to the conception of smart objects that redefine traditional user interfaces and challenge design paradigms and norms, often anticipating and setting trends before they become widespread in the industry.

References

1. R. Aernoudt, Incubators: tool for entrepreneurship? Small Bus. Econ. **23**(2), 127–135 (2004). https://doi.org/10.1023/B:SBEJ.0000027665.54173.23
2. P. Aragonés-Beltrán, R. Poveda-Bautista, F. Jiménez-Sáez, An in-depth analysis of a TTO's objectives alignment within the university strategy: an ANP-based approach. J. Eng. Technol. Manag. JET-M **44**, 19–43 (2017). https://doi.org/10.1016/j.jengtecman.2017.03.002
3. D. Battaglia, P. Landoni, F. Rizzitelli, Organizational structures for external growth of university technology transfer offices: an explorative analysis. Technol. Forecast. Soc. Chang. **123**, 45–56 (2017). https://doi.org/10.1016/j.techfore.2017.06.017
4. J. Bercovitz, M. Feldman, Entpreprenerial universities and technology transfer: a conceptual framework for understanding knowledge-based economic development. J. Technol. Transf. **31**(1), 175–188 (2006). https://doi.org/10.1007/s10961-005-5029-z
5. A. Bergek, C. Norrman, Incubator best practice: a framework. Technovation **28**(1), 20–28 (2008). https://doi.org/10.1016/j.technovation.2007.07.008
6. M. Bogers, A.-K. Zobel, A. Afuah, E. Almirall, S. Brunswicker, L. Dahlander et al., The open innovation research landscape: established perspectives and emerging themes across different levels of analysis. Ind. Innov. **24**(1), 8–40 (2017). https://doi.org/10.1080/13662716.2016.1240068
7. A. Bøllingtoft, J.P. Ulhøi, The networked business incubator—leveraging entrepreneurial agency? Spec. Issue Sci. Park. Incubators **20**(2), 265–290 (2005). https://doi.org/10.1016/j.jbusvent.2003.12.005
8. T.K. Bradshaw, T. Munroe, M. Westwind, Economic development via university-based technology transfer: strategies for non-elite universities. Int. J. Technol. Transf. Commer. **4**(3), 279–301 (2005). https://doi.org/10.1504/IJTTC.2005.006361
9. G. Buratti, G. Amoruso, F. Costa, M. Pillan, M. Rossi, O. Cordan, D.A. Dincay, PUDCAD project. towards a CAD-based game for the implementation of universal design principles in design education, in *International and Interdisciplinary Conference on Digital Environments for Education, Arts and Heritage* (Springer, 2018), pp. 154–162
10. B. Butler, P.J. Batt, Re-assessing value (co)-creation and cooperative advantage in international networks. Ind. Mark. Manage. **43**(4), 538–542 (2014). https://doi.org/10.1016/j.indmarman.2014.02.014
11. K.F. Chan, T. Lau, Assessing technology incubator programs in the science park: the good, the bad and the ugly. Technovation **25**(10), 1215–1228 (2005). https://doi.org/10.1016/j.technovation.2004.03.010
12. K.W. Chan, C.K. Yim, S.S.K. Lam, Is customer participation in value creation a double-edged sword? evidence from professional financial services across cultures. J. Mark. **74**(3), 48–64 (2010). https://doi.org/10.1509/jmkg.74.3.48
13. M.G. Colombo, M. Delmastro, How effective are technology incubators? Evidence from Italy. Res. Policy **31**(7), 1103–1122 (2002). https://doi.org/10.1016/S0048-7333(01)00178-0
14. J. Coupet, Y. Ba, Benchmarking university technology transfer performance with external research funding: a stochastic frontier analysis. J. Technol. Transf. **47**(2), 605–620 (2022). https://doi.org/10.1007/s10961-021-09856-3
15. F.A. Csaszar, An efficient frontier in organization design: organizational structure as a determinant of exploration and exploitation. Organ. Sci. **24**(4), 1083–1101 (2013). https://doi.org/10.1287/orsc.1120.0784

16. K. Debackere, R. Veugelers, The role of academic technology transfer organizations in improving industry science links. Univ.-Based Technol. Initiat. **34**(3), 321–342 (2005). https://doi.org/10.1016/j.respol.2004.12.003
17. E. Enkel, O. Gassmann, H. Chesbrough, Open R&D and open innovation: exploring the phenomenon. R and D Manag. **39**(4), 311–316 (2009). https://doi.org/10.1111/j.1467-9310.2009.00570.x
18. N.J. Foss, J. Lyngsie, S.A. Zahra, Organizational design correlates of entrepreneurship: the roles of decentralization and formalization for opportunity discovery and realization. Strateg. Organ. **13**(1), 32–60 (2015). https://doi.org/10.1177/1476127014561944
19. A. Geuna, A. Muscio, The governance of university knowledge transfer: a critical review of the literature. Minerva **47**(1), 93–114 (2009). https://doi.org/10.1007/s11024-009-9118-2
20. S. Gonzalez, D. Bettiga, J. Shao, et al., Crowdfunding: a new meaning for fundraising & user innovation, in *19th DMI* (2014), pp. 2683–2709
21. M. Good, M. Knockaert, B. Soppe, M. Wright, The technology transfer ecosystem in academia. An organizational design perspective. Technovation **82–83**, 35–50 (2019). https://doi.org/10.1016/j.technovation.2018.06.009
22. P. Gubitta, A. Tognazzo, F. Destro, Signaling in academic ventures: the role of technology transfer offices and university funds. J. Technol. Transfer. **41**(2), 368–393 (2016). https://doi.org/10.1007/s10961-015-9398-7
23. C. Gunasekara, The regional role of universities in technology transfer and economic development, in *Proceedings of the Management Futures-BAM 2004* (2004), pp. 1–28. https://eprints.qut.edu.au/1008/
24. A.N. Link, J.T. Scott, Research, science, and technology parks: vehicles for technology transfer, in *The Chicago Handbook of University Technology Transfer and Academic Entrepreneurship*, ed. by A.N. Link, D.S. Siegel, M. Wright (University of Chicago Press, Chicago, 2015), pp. 168–187. https://doi.org/10.7208/9780226178486-007
25. G.D. Markman, P.T. Gianiodis, P.H. Phan, D.B. Balkin, Innovation speed: transferring university technology to market. Creat. Spin-Off Firms Public Res. Inst.: Manag. Policy Implic. **34**(7), 1058–1075 (2005). https://doi.org/10.1016/j.respol.2005.05.007
26. M.C. Mason, S. Iacuzzi, P. Fedele, A. Garlatti, Stakeholder engagement: which arrangements for value co- creation? MM 19–43 (2020). https://doi.org/10.1431/96397
27. M. Neves, M. Franco, Academic spin-off creation: barriers and how to overcome them. R&D Manag **48**(5), 505–518 (2018). https://doi.org/10.1111/radm.12231
28. J.A. Pearce, P.C. Patel, Reaping the financial and strategic benefits of a divestiture by spin-off. Bus. Horiz. **65**(3), 291–301 (2022). https://doi.org/10.1016/j.bushor.2021.03.001
29. K. Pedersen, What can open innovation be used for and how does it create value? Gov. Inf. Q. **37**(2) (2020). https://doi.org/10.1016/j.giq.2020.101459
30. L. Peters, M. Rice, M. Sundararajan, The role of incubators in the entrepreneurial process. J. Technol. Transf. **29**(1), 83–91 (2004). https://doi.org/10.1023/B:JOTT.0000011182.82350.df
31. J.B. Powers, R&D funding sources and university technology transfer: what is stimulating universities to be more entrepreneurial? Res. High. Educ. **45**(1), 1–23 (2004). https://doi.org/10.1023/B:RIHE.0000010044.41663.a0
32. E. Rasmussen, Ø. Moen, M. Gulbrandsen, Initiatives to promote commercialization of university knowledge. Technovation **26**(4), 518–533 (2006). https://doi.org/10.1016/j.technovation.2004.11.005
33. V. Schaeffer, M. Matt, Development of academic entrepreneurship in a non-mature context: the role of the university as a hub-organisation. Entrep. Reg. Dev. **28**(9–10), 724–745 (2016). https://doi.org/10.1080/08985626.2016.1247915
34. A. Schoen, B. van Pottelsberghe de la Potterie, J. Henkel, Governance typology of universities' technology transfer processes. J. Technol. Transf. **39**(3), 435–453 (2014). https://doi.org/10.1007/s10961-012-9289-0
35. M. Sciarelli, G.C. Landi, L. Turriziani, M. Tani, Academic entrepreneurship: founding and governance determinants in university spin-off ventures. J. Technol. Transf. **46**(4), 1083–1107 (2021). https://doi.org/10.1007/s10961-020-09798-2

36. D.S. Siegel, R. Veugelers, M. Wright, Technology transfer offices and commercialization of university intellectual property: performance and policy implications. Oxf. Rev. Econ. Policy **23**(4), 640–660 (2007). https://doi.org/10.1093/oxrep/grm036
37. D.S. Siegel, D.A. Waldman, L.E. Atwater, A.N. Link, Commercial knowledge transfers from universities to firms: improving the effectiveness of university–industry collaboration. J. High Technol. Managem. Res. **14**(1), 111–133 (2003). https://doi.org/10.1016/S1047-8310(03)00007-5
38. A. Smith, A. Stirling, F. Berkhout, The governance of sustainable socio-technical transitions. Res. Policy **34**(10), 1491–1510 (2005). https://doi.org/10.1016/j.respol.2005.07.005

Chapter 4
A Pathway to Innovation. Design-Driven and User-Centric

Abstract In this chapter, the focus is on Thingk's approach to innovation, characterised by its design-driven and user-centric methodology in developing smart home products. The discourse unfolds through a critical analysis of a three-phase innovation process: Research, Design, and Development. It uses the case study of *Slab!*, an IoT device serving as a kitchen scale or timer, to illustrate how complex functionalities can be seamlessly integrated into products with minimalistic aesthetics. Challenges and shortcomings encountered during the process are specifically addressed.

Keywords Design-driven innovation · Human-centric design · User engagement · Prototyping · Iterative design

4.1 Design-Driven Innovation and a User-Centred Approach

Designing innovative products in a context where digital technology is becoming more accessible, human-centric, and empowering, reshaping user habits and industrial patterns [11] requires new approaches and pathways to innovation. The ongoing progress nurtured a shift from mere functionality to desirability, with users expecting immediate, simple, and almost invisible interactions that enable more natural and continuous user experiences (UX), thereby reducing the cognitive load [21, 23]. As a consequence, the field of design itself is evolving, requiring a blend of multidisciplinary knowledge. This includes integrating insights from sociology, anthropology, ethnography, and marketing into the design process.

Within this research and innovation landscape, Thingk designs apparently traditional artefacts with 'superpowers'. The spin-off experiments with IoT and emerging technologies in Smart Home products, designing seemingly static artefacts that appear simple and minimal (in form) but are technologically enhanced (in function). This involves embedding interfaces that are subtle, dissolved, and invisible until interaction. Thingk's exploration in this challenging direction led to the cross-pollination of disappearing interfaces with semantically connected morphological

U. Tolino and I. Mariani, *Design Behind Interaction*,
PoliMI SpringerBriefs, https://doi.org/10.1007/978-3-031-67416-7_4

forms. The resulting products, characterised by an analog appearance but smart capabilities, challenge traditional aesthetic-functional conventions, necessitating a critical examination of how users perceive and interact with these artefacts. As such, Thingk's products aim at challenging and reshaping user habits through design [31].

In light of needs such as continuous technological updates and a market analysis aimed at anticipating trends, the discipline of design is called upon to evolve and reinvent itself, and open up to diversity. This evolution demands a hybridisation of knowledge that entails a revision of the traditional design process, moving towards greater inclusiveness of interdisciplinary and transdisciplinary perspectives and scholarship. The current context underscores and rewards the opportunities that arise from intertwining research and development in the fields of design, engineering, and other disciplines, in order to design products that combine responses to needs in terms of digital performance with an elegant and discreet aesthetic, based on refined research into materials and interactions.

Moreover, the design of smart objects and user interfaces that challenge and question traditional design paradigms underscores the importance of engaging users and stakeholders. Introducing innovative interaction methods that diverge from standardised and established ones necessitates involving users and stakeholders, including manufacturers, into the design loop, ensuring their needs, challenges, and perspectives are considered. As experts of their needs [17, 18], their expertise is often complementary to that of researchers and designers, leading to insights that can largely inform design decisions. For these reasons, their involvement requires being strategically and meaningfully planned throughout the design process. By engaging users and stakeholders from the outset and consistently throughout the development, designers can ensure that the resulting products meet both practical and performance standards while addressing real-world usability and accessibility issues, hence resonating with the aesthetic and functional needs of end-users. Such collaboration fosters ongoing feedback and adaptation, which nurtures the improvement of aesthetics and functionalities of smart objects and interfaces to better meet the dynamic needs of users and adapt to the evolving technological landscape.

To address these fundamental aspects and foster a holistic approach to product development, the spin-off has built around a team with complementary backgrounds that span multiple areas of design and engineering, integrating expertise from management, sociology, anthropology, ethnography, and marketing.

In light of this reasoning, this chapter illustrates our design-driven innovation and user-centred approach, showing how this complexity led to defining a specific design process. The adoption of a design-driven innovation approach [24, 26] ensures to anticipate user needs and market demands [6], incorporating new technologies into distinctive products. This approach is complemented by a user-centred perspective that engages users and stakeholders in various stages of design [1, 10, 14, 32]. The integration of the two approaches situates the user at the heart of the design and development process, ensuring that products are not only technologically advanced but also intuitively align with user needs and expectations, while aiming at innovating how users can interact with products and systems.

The process unfolds as a **pathway to innovation** consisting of three main phases, critically exploring their benefits and challenges.

The first phase, Research, involves comprehensive user research, technology and innovation benchmarking, market product analysis, and aesthetic exploration. It allows for anticipating user needs, aligning with market trends, and exploiting technological advancements [16, 20]. The subsequent Design phase consists of an iterative cycle of design, prototyping, user testing with mixed methods, and data analysis, all pivotal for informing and orienting further implementations of the artefact, based on user interaction and feedback. The third phase, Development, follows a similar iterative pattern, focusing on production, distribution, usage data and user feedback, including insights gathered from social media interactions and product reviews, and continual product improvement to enhance the overall user experience, especially through software updates while maintaining the hardware.

The process depicts the progression along the **Technology Readiness Level (TRL) scale** (Mankins and others 1995), simultaneously showing the shift from two distinct types of innovation [15, 25, 26]. The Research and Design phases relate to radical innovation, introducing significant advancements and novel concepts. Conversely, the Development phase aligns to an incremental innovation, mainly focusing on continuous improvement and refinement of the existing technological framework for improving functions. Software enhancements allow for mitigating or postponing the risks associated with obsolescence, contributing to the longevity and relevance of the technology.

These phases, here just summarised, are elaborated and detailed in the next paragraphs. To explain the process, *Slab!* serves as a case study. Its overall process is discussed, including its user testing through a crowdfunding for market interest and user feedback, underscoring the importance of involving users in improving the product in terms of experience and functionality [22, 27].

4.2 The Design Process and Its Phases

Our design-driven innovation, with a user-centred approach, unfolds within a design process aimed at conceiving and creating objects with mutable meanings, whose language cannot be taken for granted. The objects we produce appear analog but are actually technologically enhanced (smart). They are also deliberately and designerly deprived of an explicit interface, and thus can be considered to have a dual semantic level, which varies depending on whether the interface is silent because the object is not used, or visible when the object is interacted with.

The design process underpinning the research and development of these objects reflects how functions hidden beneath the surface emerge as needed, and the implications thereof. The challenge of making interaction intuitive, despite its levels of complexity, is on the ground of a rich and articulated process, where innovation stems from iterative cycles of research, experimentation, user testing, and implementation.

The process is structured into three broad phases. In each phase, the multi- and transdisciplinary skills and knowledge of the spin-off play a fundamental role in fostering creativity and innovative practices, drawing simultaneously from the fields of design and engineering and integrating them with scholarship from humanistic disciplines. In doing so, the role of marketing also becomes apparent, which has the important task of strategically incorporating market signals into the research of new products. Furthermore, whether it is during the analysis of the market segment and prospect desiderata, for an initial interaction with a project idea or prototype, or for testing and data collection on use: in various ways and to different extents, users are involved throughout the entire process. Stakeholders are also involved as key actors of the ecosystem, particularly those manufacturers who have distinguished themselves in their sectors for particular research and innovation. Their involvement brings productive innovation and experimentation with new materials to the process, derived from experience in specific sectors.

The first phase concerns research, and the second to design. These two phases are related to **radical innovation in meaning**, which occurs through conceptual innovation that can pertain to the product and its type. In our case, it involves introducing technology into an object in such a way as to rethink its meaning (see Chap. 5), to meet a market need in an unexpected manner. The third phase concerns development, and includes a type of **daily innovation** that is more incremental and tied to the refinement of specific aspects and qualities to improve the product. Throughout this process, the advancement in the steps of the Technology Readiness Level (TRL) is highlighted.

The design process map (Fig. 4.1) associates phases with the advancement of the technological product's maturity. These TRL stages are visualised as a journey from conception to market readiness, demonstrating how a project evolves from basic research through to a final product that is ready for consumer use. It also shows how feedback loops are integral, particularly user engagement which helps refine and qualify the product at various stages, ensuring that it is ready for real-world application.

4.2.1 Research

The research phase unfolds from the recognition of technology's role in the development of competitive products. The computational power that objects can possess and their smart capability provide opportunities that include additional functionalities and greater reliability. Designing products with variable functionalities necessitates a re-evaluation of value chains, which in turn impacts future market dynamics [16]. The value of research is crucial for generating a competitive edge. The strategic importance of research lies in its ability to produce a distinct competitive advantage in such a transformative landscape.

The research phase begins with a **trend analysis**, which delves into current market trends and identifies technological advancements that are both intriguing and hold

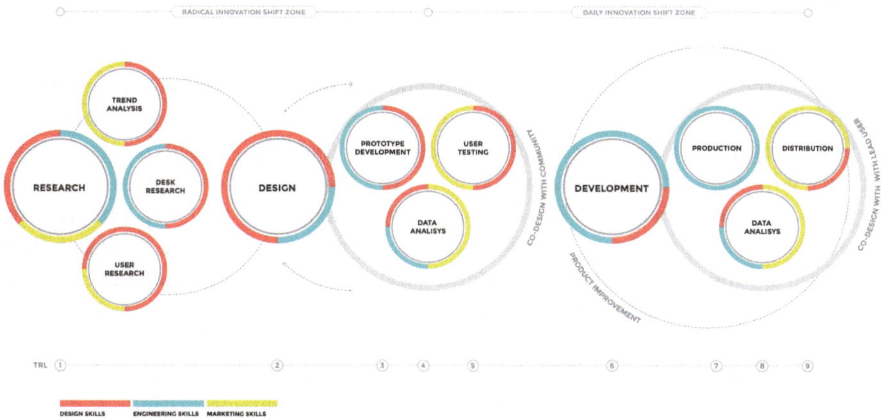

Fig. 4.1 Iterative design process with phases, domains, typologies of innovation and TRL

potential promise. This phase consistently analyses trends in technology and lifestyle to pinpoint promising directions. The analysis aims to explore experimentation and inspiration in terms of functionality and usability, language and meaning, with an awareness that, by vocation, we position ourselves in an area of friction where the interface tends to disappear to merge with the object. Within the strategic areas of interest we've identified, we trace the state of the art, and with it the boundaries of current experimentation, to identify areas where we can conduct explorations and develop products. In doing so, advancements in electronics, interaction design, and material research—especially looking at the potential to merge traditional aesthetics with smart materials that rewrite their meaning and functions—are considered.

In this subsequent stage, the previously identified trends and technological advancements are analysed in-depth with a comprehensive desk research. This involves a review of existing literature and a detailed examination of the state of the art, aiming to consolidate theoretical foundations as well as practical knowledge. The context for our explorations is that described in Chap. 2. In this landscape, industry studies indicate a trend for technology to dissolve into the body, the environment, and the cloud [20], fueling the search for interactions that benefit from this possibility [2, 7, 11, 19]. This trends are further explored through an **analysis of the literature (desk research)**, which shows significant experimentation in this regard: from interactive objects where aesthetics make the interaction tangible, to objects with natural interfaces that rely on gesture control to manage commands, to artefacts that hide their function behind a silent interface that emerges only when needed. A prominent aspect then concerns the integration of the IoT technological component that rethinks objects as extended and interconnected ecosystems, capable of interpreting contexts, communicating with other objects and systems, in a view of progressive simplification and often even prediction.

When dealing with interaction and communication, materials play a central role, and the concept of **materiality** in particular [29, 30]. Developed in the semantic dimension of manual interaction with artefacts, materiality and the role of tactile experience take on new meanings. The surfaces of the artefacts become the transition point from the physical to the digital world, and the information that derives from the tangible aspects of objects extends from the analog context to convey immateriality.

Following the desk research, **user research** is undertaken. This stage engages directly with the end-users and stakeholders to gather empirical insights. Through methods such as interviews, focus groups, and observation, this step uncovers the needs, behaviours, and preferences of the target audience as well as possibilities coming from the experimentation undertaken by manufacturers and relevant stakeholders in the ecosystem. It ensures that the project is firmly rooted in actual user requirements and experiences, allowing for a design that is both relevant and resonant with its intended users. This stage is an essential link between the theoretical underpinnings of the project and its practical, user-oriented application.

Overall, the initial Research phase mirrors TRL 1 (Basic principles observed), concerning the exploration of multiple possibilities, leading to the generation of initial concept ideas that are both abstract and boundless. This is the stage where innovation starts to bud.

4.2.2 Design

The outcomes of the research phase guide the design process, benefiting from cross-pollination within an interdisciplinary team whose knowledge and know-how stretch across diverse areas of design and engineering. Our team tends to explore and emphasise particularly alternative directions for interactions capable of eliciting unconventional user experiences that are surprising and sophisticated.

Embracing a tendency towards technological hybridization, our design approach favours a merge between the analog dimension and the immateriality of the digital: we prioritise the use of natural materials combined with the search for essential forms. This results in investigating the haptic perception of artefacts, which in our specific case fuels constant cognitive friction. Following a pursuit of apparent simplicity, this haptic dimension of the object directs user expectations towards certain interactions, based on usage habits reinforced by the shape and materiality of the artefacts. However, the simplicity inferred from the appearance of the products intentionally conceals complexity due to unexpected intelligence that emerges from the surface when needed. This **semplexity** develops from the union of archetypal, geometric, and minimal shapes, married to sophisticated materials made smart through the integration of technology. Examples include a milled wooden cutting board that actually conceals being a smart scale/timer, a desk pad that functions as a Wi-Fi charger in handcrafted leather, and a polished stone containing a wireless data storage system.

We follow the principles of design-driven innovation [24, 26], fueled by thorough and distributed research in a context populated by expters from different disciplinary domains. These interpreters are capable of capturing cultural and technological trends and translating them into significant and creative contributions that take the form of new objects with unexpected functions. For the prototyping phase, we embrace a user-centred approach [1, 32]. Following the creation of the **prototype** is the experimentation with users, which we conduct using the ethnographic method of participant observation, focusing on capturing evidence and insights about how users respond to interaction with objects that disguise their true nature, combining physical and digital aspects. Acting on the form and introducing technology that hacks the meanings of objects requires careful evaluation of how users perceive, interact, and "translate" the artefact. To obtain clear and detailed information, we prefer to conduct extended observation over time, based on real activities and not necessarily guided tasks, with participation in the activities of the observed users.

Despite typically associated with technological dimensions, radical innovation [25, 26] embraces a far broader domain. Although less frequent, radical changes can also occur within the sphere of meaning. Designing through **meaning-driven innovation** [15] requires careful observation of the cultural fabric, identifying subtle yet unexpressed socio-cultural dynamics and potential or ongoing changes.

Observing the context demands acute attention to demographic shifts and lifestyle changes [15]. In specific instances, it is indeed considered counterproductive to be guided by users during the research and design phases, considering that profound innovation is unlikely to develop from those with a fixed understanding of function and meaning. Conversely, during development and testing phases, user involvement is advantageous to gauge reception to radical innovations that have shifted meanings. Performing this analysis during the prototyping phase allows for timely action on any discrepancies or misunderstandings, leading to new cycles of implementation and testing until issues are resolved. Early observation during prototyping can identify recurring usability issues with the object or its interface—a concept that becomes clearer when applied to a case study, as will be illustrated below.

Gaining a clear and reliable evaluation of the product in terms of usability, aesthetics, and especially materiality, proportions, forms, and their affordances, as well as the user experience, including symbolic and emotional aspects, is of utmost importance. To optimise this evaluation, when possible, we involve communities of prosumers already engaged in the qualitative research phase. Through creative brainstorming sessions, focus groups, and interviews, we aim to understand how the envisioned innovation might be received and perceived, favouring the inclusion of designers for critique and analysis. User involvement then pauses during the initial design phases and resumes in the product testing and validation phase, where we conduct focused group discussions and targeted interviews. We believe that creative activities do not require user interaction as they rarely contribute to anticipating potential innovations [6]. Users tend to be bound to the present and its possibilities, a trait that hinders the ability to envision nonexistent applications. The inclusion of prosumers, however, becomes crucial in the testing and validation phase of the iterative process [22].

The Design phase parallels TRL 2 (Technology concept formulated), where the previously abstract ideas begin to take shape. At this juncture, the focus is on formulating technology concepts into tangible designs concepts. The design process translates the speculative into the concrete, setting the stage for the development of IoT products that will meet emerging market trends and user needs. The Prototype Development reflects TRL 3 (Experimental proof of concept). Materialising the concept through prototyping, this stage is an experimental playground where the proof of concept is validated within a controlled setting, that of the spin-off. This stage critically signifies the first tangible intersection of design theory and technological application, establishing a foundation for user-centred IoT products.

The design undergoing User Testing and Data Analysis relates to TRL 4 (Technology validated in lab) and TRL 5 (Technology validated in relevant environment), where the functionality and usability are observed both in the spin-off setting and in situ, with specific experimentation in environments that mimic where the product will be ultimately deployed. Here, the iterative process of validation ensures that the IoT product is not just a mere technological contribution but a solution finely tuned to the user's context.

4.2.3 Development

In developing our products, we aim to adopt flexible and reconfigurable production logics. We are aware of the rapid obsolescence that features electronic and technological components (both hardware and especially software), and the consequent need to foresee agile update models with low impact on the production chain. This approach allows us to guarantee competitiveness along with a high capacity to adapt to the market.

Designing with a perspective of flexibility rather than standardisation implies engineering the shape and electronics of products to be predisposed to possible structural changes [3]. Consequently, we adopt a logic of counter-economies of scale, which, however, allows us to develop objects from an engineering and design point of view that can reconfigure according to needs. For example, hardware technological updates can trigger different levels of needs that imply revising aspects such as shape and the simultaneous housing of multiple elements in reduced spaces. Conversely, software updates do not carry these types of implications, although they can have more complex impacts at the systemic level. Especially when it comes to smart objects that employ IoT technologies and are connected in a larger system of elements, the reasoning presents additional complications related to post-update compatibility.

After completing the development and engineering of electronic components, production follows two different manufacturing streams. The body of the product is made by involving highly-specialised manufacturers in the processing of refined materials that constitute the object, often involved already during the research process to explore materials and new production methods and processes. At this point, a series of experiments is started to create study samples aimed at testing the different options. In parallel, companies operating in the electronics sector are identified and engaged for the production of technological components.

Once the optimal solutions are identified, a first limited series of artefacts is produced, promoted, and tested. At this point in the process, a community of early-adopters is identified, involved, and observed, which allows us to verify how an object whose meaning has been redefined is perceived while it is in use, directly with the users [4, 9, 12], in terms of both usability and intelligibility. What we collect from this activity triggers a precise redesign that feeds the implementation.

The Development phase is where real-world application becomes the proving ground, being related to TRL 6 (Technology demonstrated in relevant environment). In-situ testing with users and stakeholders provides invaluable insights, ensuring the product is robust enough to handle the varied and unpredictable nature of the real world. This phase is crucial for IoT products, where the interplay between the digital and physical realms must be seamless.

Approaching Production, correlating with TRL 7 (System prototype demonstration in operational environment), the system prototype undergoes rigorous testing with early adopters in the actual operational environment. This is the phase where the prototype is more than a prototype. It functions as a complete working system, ready for the rigours of daily use.

Finally, towards the end of the journey lies TRL 8 (System complete and qualified) and TRL 9 (Actual system proven in operational environment), relating to Distribution and beyond, where the IoT product has been thoroughly vetted and proven in the operational environment. At this stage, the product is a market-ready IoT solution, and therefore enters the market.

4.2.4 *Slab!*: The Process Applied

To bring this process to life, we discuss the specific case of *Slab!*, a product by Thingk designed with an intuitive, gestural interaction concept. We critically analyse how the interface was perceived by early adopters involved during the design process, detailing the changes that insights gained have made on the final product.

Through the critical analysis of *Slab!*, a smart kitchen scale that also functions as a cutting board and digital timer (Figs. 4.2 and 4.3), we explore how users perceive and interact with the product.

Slab! is the first object designed by the spin-off. As described in Chap. 3, it is a kitchen scale, a digital timer, and an IoT device prepared to communicate with other smart devices. The smart kitchen scale is designed to create a user experience that overlays the physical sphere with the digital one. It appears to be a cutting board; however, when the user manipulates the object by changing its position or orientation, information emerges from the wooden surface revealing its true functions as a scale and timer.

However, to achieve the final aesthetic form, *Slab!* underwent a series of tests and implementations, which involved users throughout the entire design process.

The design process of this case is contextualised within the collaborative process determined by the crowdfunding campaign carried out to support the conceptual

Fig. 4.2 Main functions of *Slab!,* described through frames from a video showing the smart scale in use

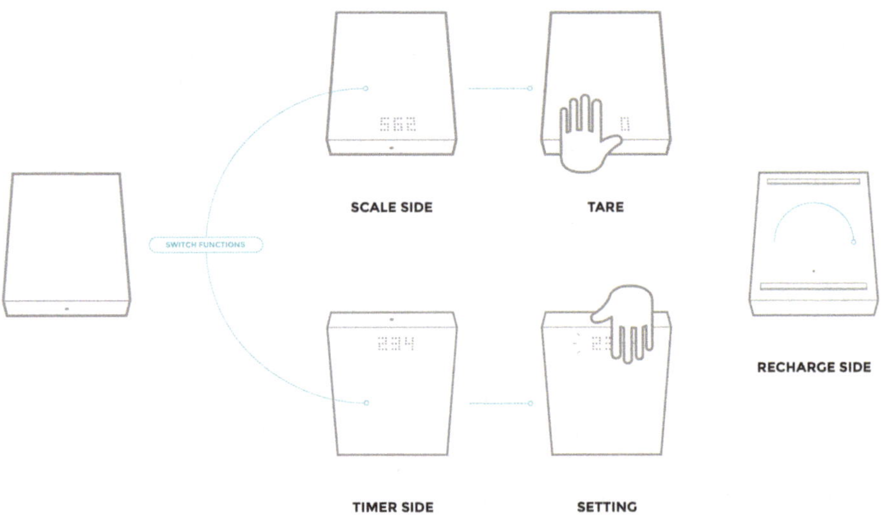

Fig. 4.3 Functions of *Slab!,* activated through gestural interaction or manipulation

idea behind this product. Adhering to the crowdfunding timelines, some phases—research and design of the prototype—were condensed and accelerated, as shown in Fig. 4.4. Moreover, throughout the design journey, users were placed at the centre of the design process we describe next.

Research. Based on the analysis of the state of the art and technological possibilities, the product idea emerged from localised information in space and the concept of making it responsive to the manipulation and orientation of the object itself. This product pursues an ideal of natural and intuitive interaction that is based on the principles of calm technology [5, 28], combined with a desire to make not only

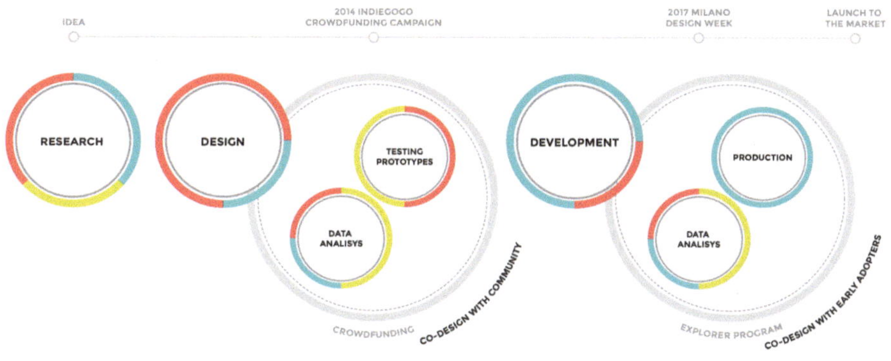

Fig. 4.4 The iterative design process of *Slab!*

the information but also the function itself tied to the concept of being situated and contextualised. The underlying concept of Slab! is that it appears to be a cutting board, but a simple press or change of orientation reveals its real functions. The entire research setup corresponds to that of this volume. The initial user research phase mainly involved early adopters of smart objects and stakeholders in the home appliances sector.

Design. The first concept of *Slab!* was created in 2014, aiming to design an object that is aesthetically minimal and technologically enhanced. Slab! employs a user interface designed to emerge when the product is interacted with and to remain silent when at rest. To finance its development while simultaneously reaching a broad and active user community that could assist us in the task, we launched a crowdfunding campaign on the Indiegogo platform. This campaign allowed us to reach and build a significant community of prosumers [8] with whom we conducted experiments to explore the potentials and implications of a product with a design that is naturally ambiguous, while the product was still in its prototype phase. The crowdfunding campaign is used to consistently engage users with the first working prototype of *Slab!* and inform its redesign.

Following the crowdfunding campaign that began in October 2014, over 200 *Slab!* units were shipped to an international community across 19 countries. For six months, we engaged with those who participated in the crowdfunding and received the smart scale to understand their initial reactions to the product, how they interpreted it, and the functions it communicated to them. We asked them to share their experiences and opinions, which provided diverse perspectives on how the product was used and understood. The feedback revealed insights into how the embedded technology, once disclosed, informed users about the product's functionality.

Specifically, the feedback gathered guided the redesign of five closely related aspects, each triggering a series of implications (Fig. 4.5):

- Orientation and its relationship with functions (aesthetics)
- Position of the timer (functionality)
- Display position of the scale (aesthetics)

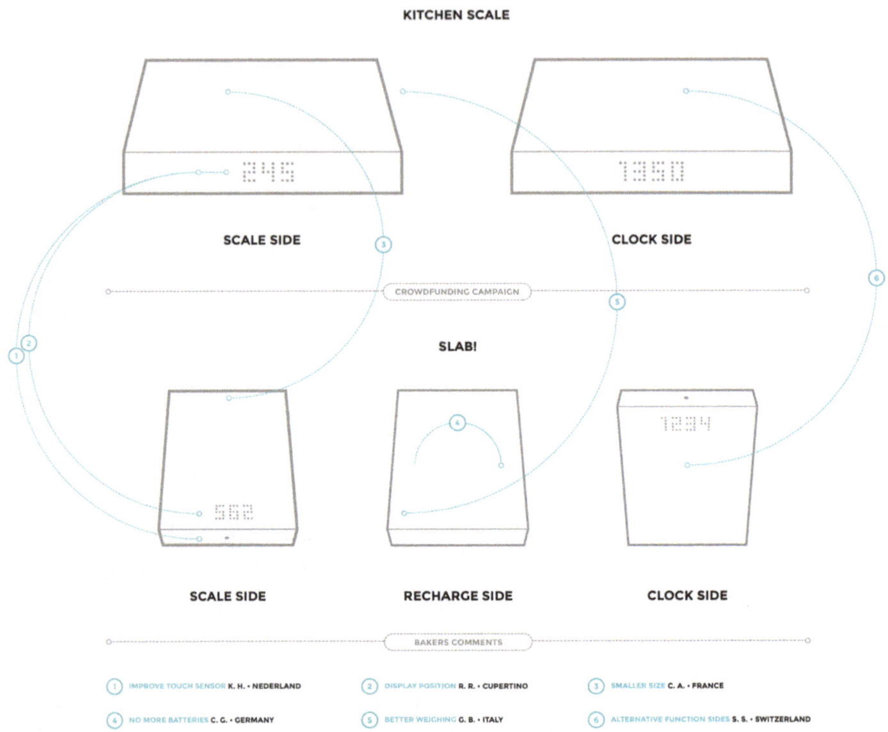

Fig. 4.5 Changes implemented based on feedback from the users engaged through the crowdfunding campaign

- Size of the object (aesthetics)
- Tare button (functionality and aesthetics)
- Charging mode (functionality).

In summary, the feedback raised questions about user interaction due to its minimalist aesthetics, revealing communication issues, especially regarding the object's operational orientation (Fig. 4.6). To address this, a graphic mark (the Thingk logo) was added to indicate the touch-sensitive area for setting the scale or timer, depending on orientation. Additionally, the LCD display was relocated to the top side for better visibility and to clarify operational orientation, significantly enhancing usability. Observations on the object's size led to a 25% reduction in its dimensions to better fit its environment. The project culminated with the introduction of wireless charging, adopting the Qi standard to align with emerging trends at the time.

The feedback from the campaign led to iterative modifications based on a deep analysis of user interactions, leading to a revised prototype (Fig. 4.7).

Development. The revised prototype is the object of further testing with a selected group of early adopters through an explorer program that gave them access to the

Fig. 4.6 The prototype of the kitchen scale delivered to the crowdfunding community

Fig. 4.7 *Slab!* with the display repositioned after the analysis phase carried out with crowdfunding campaign users

revised version of *Slab!*. Adjustments included repositioning the LED display, modifying the functions based on the orientation, and resizing the device which was perceived as too large. Observations from a smaller sample that interacted with the implemented product allowed us to validate our design decisions. Generally, the feedback collection provided a clear understanding of how our prototype and later our product were used in line with our expectations, highlighting the critical role of data in understanding and evaluating real user interactions.

This analysis goes hand in hand with studying perceptions of design aspects such as materials, ergonomics, and usability. For *Slab!*, comments mainly focused on functionality and aesthetics—the former regarding the extent to which the artefact was capable of performing desired or expected operations, and the latter indicating how well the artefact communicated its functions (affordance). Feedback clearly showed how the product's minimalistic aesthetics, coupled with its enhanced nature, challenged established norms.

In the development of *Slab!*, flexibility was prioritised to accommodate updates to its technological components. Since its inception, hardware updates have necessitated changes to the internal structure of the casing, particularly the sizing of electronic compartments. Opting for artisanal construction methods over standardised mass production has facilitated these hardware updates, allowing for quick and precise reconfigurations without economic impact on the production chain. The craftsmanship has achieved an exceptionally thin wood casing, ensuring the LCD is visible through the surface. A designer community participated in a second phase, analysing potential technological implementations, which turned out to be too nascent for integration.

* * *

The design process applied to *Slab!* highlights the integration of smart technology in objects, attention to emerging usage patterns, and adaptation of products to a constantly evolving context. Designing with these considerations allows for responsiveness to opportunities and challenges arising from questioning established aesthetic-functional conventions of household objects, thereby altering their interpretative processes.

This approach underscores a dynamic interface strategy, adaptable and expressive of various functions depending on the usage context, reflecting the ongoing experiments aimed at merging multiple meanings into a single object. The overall process exemplifies the importance of a flexible, iterative design approach in responding to rapid technological advancements and market changes, signalling a paradigm shift in design artefacts that meld smart and emerging technologies with minimal aesthetics.

The overall process emphasises the significance of a flexible, iterative design approach in responding to rapid technological advancements and market changes, signifying a paradigm shift in design artefacts that integrate smart and emerging technology with minimal aesthetics.

References

1. C. Abras, D. Maloney-Krichmar, J. Preece, User-centered design. Berks. Encycl. Hum.-Comput. Interact.; Thousand Oaks: Sage Publications, **37**(4), 445–456 (2004)
2. S. Ballmer, *CES 2010: A Transforming Trend — The Natural User Interface*. (2010). www.huffingtonpost.com/steve-ballmer/ces-2010-a-transforming-t_b_416598.html
3. M. Bianchini, S. Maffei, Microproduction everywhere: Defining the boundaries of the emerging new distributed microproduction socio-technical paradigm, in *Social Frontiers: The Next Edge of Social Innovation Research* (2013), pp. 1–21
4. M. Bogers, A. Afuah, B. Bastian, Users as innovators: a review, critique, and future research directions. J. Manag. **36**(4), 857–875 (2010). https://doi.org/10.1177/0149206309353944
5. A. Case, *Calm Technology: Principles and Patterns for Non-intrusive Design* (O'Reilly Media, Sebastopol, CA, 2015)
6. C. Dell'Era, N. Altuna, S. Magistretti, R. Verganti, Discovering quiescent meanings in technologies: exploring the design management practices that support the development of technology epiphanies. Technol. Anal. Strat. Manag. **29**(2), 149–166 (2017). https://doi.org/10.1080/09537325.2016.1210785
7. E. Giaccardi, Designing the connected everyday. Interactions **22**(1), 26–31 (2015). https://doi.org/10.1145/2692982
8. S. Gonzalez, D. Bettiga, J. Shao, et al., Crowdfunding: a new meaning for fundraising & user innovation, in *19th DMI* (2014), pp. 2683–2709
9. C. Herstatt, E. Von Hippel, From experience: developing new product concepts via the lead user method: a case study in a "low-tech" field. J. Prod. Innov. Manag. **9**(3), 213–221 (1992)
10. C. Kraft, *User Experience Innovation: User Centered Design that Works* (A Press, New York, NY, 2012)
11. M. Kuniavsky, *Smart Things, Ubiquitous Computing User Experience Design* (Elsevier, Burlington, MA, 2010)
12. C. Lüthje, C. Herstatt, The lead user method: an outline of empirical findings and issues for future research. R&D Manag. **34**(5), 553–568 (2004)
13. J.C. Mankins, et al., Technology readiness levels. White Paper, April, 6 (1995)
14. M.C. Mason, S. Iacuzzi, P. Fedele, A. Garlatti, Stakeholder engagement: which arrangements for value co-creation? MM 19–43 (2020). https://doi.org/10.1431/96397
15. D.A. Norman, R. Verganti, Incremental and radical innovation: design research vs. technology and meaning change. Des. Issues **30**(1), 78–96 (2014)
16. M.E. Porter, J.E. Heppelmann, How smart, connected products are transforming competition. Harv. Bus. Rev. **92**(11), 64–88 (2014)
17. E.B.-N. Sanders, P.J. Stappers, Co-creation and the new landscapes of design. CoDesign **4**(1), 5–18 (2008). https://doi.org/10.1080/15710880701875068
18. L. Sanders, P.J. Stappers, From designing to co-designing to collective dreaming: three slices in time. Interactions **21**(6), 24–33 (2014). https://doi.org/10.1145/2670616
19. O. Shaer, E. Hornecker, et al., Tangible user interfaces: past, present, and future directions. Found. Trends® Hum.–Comput. Interact. **3**(1–2), 4–137 (2010)
20. T. Sutton, Invisible applications. UX design without GUIs. Mediu. (J. Electron.) (2014). https://medium.com/@thomas_thinks/invisible-applications-10501f6cfa5
21. J. Sweller, Cognitive load theory, learning difficulty, and instructional design. Learn. Instr. **4**(4), 295–312 (1994)
22. U. Tolino, I. Mariani, Do You think what i think? Strategic ways to design product-human conversation. Strat. Des. Res. J. **11**(3), 254–262 (2018). https://doi.org/10.4013/sdrj.2018.113.10
23. S. Turkle, *Alone Together: Why We Expect More from Technology and Less from Each Other* (Basic Books, New York, NY, 2011)
24. J. Utterback, B. Vedin, E. Alvarez, S. Ekman, S. Walsh Sanderson, B. Tether, R. Verganti, *Design-Inspired Innovation* (World Scientific Publishing, Singapore, 2006)

25. R. Verganti, Design, meanings, and radical innovation: a metamodel and a research agenda. J. Prod. Innov. Manag. **25**(5), 436–456 (2008)
26. R. Verganti, *Design-Driven Innovation—Changing the Rules of Competition by Radically Innovating What Things Mean* (Harvard Business Press, Boston, MA, 2009)
27. I. Vitali, V. Arquilla, U. Tolino, A design perspective for IoT products. A case study of the design of a smart product and a smart company following a crowdfunding campaign. Des. J. **20**(sup1), S2592–S2604 (2017). https://doi.org/10.1080/14606925.2017.1352770
28. M. Weiser, J.S. Brown, Designing calm technology. PowerGrid J. **1**(1), 75–85 (1996)
29. M. Wiberg, H. Ishii, P. Dourish, D. Rosner, A. Vallgårda, P. Sundström, et al., Material interactions: from atoms & bits to entangled practices, in *CHI'12 Extended Abstracts on Human Factors in Computing Systems* (ACM, 2012), pp. 1147–1150
30. M. Wiberg, E. Robles, Computational compositions: aesthetics, materials, and interaction design. Int. J. Des. **4**(2), 65–76 (2010)
31. F.R. Wilson, *The Hand: How its Use Shapes the Brain, Language, and Human Culture* (Vintage, New York, NY, 1999)
32. P. Wright, J. McCarthy, Experience-centered design: designers, users, and communities in dialogue. Synth. Lect. Hum.-Cent. Inform. **3**(1), 1–123 (2010)

Chapter 5
Semantic Reconfigurations. Exploring Implications on Agency and Affordances

Abstract This chapter explores the implications derived from including embedded technology and interfaces on object design. Particularly, it dives into how this approach leads to a semantic reconfiguration of the meaning attributed to such objects, hence influencing the communication of functionalities and challenges traditional design paradigms. It critically examines the effects on agency and affordance in the context of objects that do not explicitly reveal their interaction modes. Anchored in a postphenomenological discourse, the chapter ultimately unfolds the complexities of designing objects with hidden or misleading interfaces.

Keywords Design paradigms · Meanings · Agency · Affordance · User experience

5.1 Introduction: Making Sense of Things

Design as a discipline has continually evolved, serving as a bridge between creativity and functionality [14]. It is a discipline that not only crafts objects but also embeds meaning and functionality into them, influencing our daily experiences and interactions [2, 24, 31]. This evolution becomes increasingly relevant in the context of emerging technologies, where the traditional boundaries of design are challenged and expanded [30]. Before progressing in this direction and exploring how design converges with technology to create products that not only serve functional purposes but also communicate, interact, and adapt in ways previously unimagined, we consider relevant ground the discussion in the foundational definitions of design. In conceptualising its meaning, [26] describes Design as the attribution of sense to things (de + signare), being related to defining the form and function of objects, systems, or interactions. Complementing this, [17] broadens the notion to the shaping of our surroundings, whether in the context of products, systems, or environments, to meet needs enhancing functionality, aesthetics, and usability. This inclusive perspective views design as not only crafting attractive objects but as solutions that improve human interaction with the world. Such a comprehensive view

U. Tolino and I. Mariani, *Design Behind Interaction*,
PoliMI SpringerBriefs, https://doi.org/10.1007/978-3-031-67416-7_5

of design is particularly pertinent to the exploration of embedded technology and interfaces in object design where the aesthetic-functional conventions of traditional products are reimagined.

From the encounter of these definitions, it becomes clear that the objects *should*, by design, communicate their function through their appearance, providing users with clues that suggest their correct operation ([34], p. 39). Aesthetics play a key role in conveying this information, and the affordances of an object suggest its use, giving the user guidance and indications on how to manipulate the artefact to achieve the desired effect [36]. Within this landscape, when it comes to objects with a digital nature, the aesthetics of the object are enriched with an interface, contributing to shaping and influencing users' behaviours and actions.

What has just been described represents how the design practice *should be*. It becomes interesting to veer towards what happens when these conventions are challenged—especially regarding the communicative capacity of products. Redesigning meanings is per se a highly complex aspect. However, it increases in complacence when it takes place within a design-driven innovation process which exploits technological opportunities and open new possibilities for experimentation [40, 44], making the form-function relationship of products no longer so familiar [23]. When objects alter their meaning, their ability to communicate their affordances becomes far from being taken for granted, and this comes with significant implications.

5.1.1 Challenging Conventions by Design

The interface acts as a holistic system that engages in dialogue with the user [25], conveying product, service, or system functionalities, informing users about potential uses (affordances) [34]. This interactive and performative dimension is what transforms objects into responsive entities [3], and usability is just one aspect of a larger set of characteristics contributing to this dialogue. In this framework, technology not only reshapes our perception but also redefines possible actions, thus influencing behavioural outcomes. Affordances, intertwined with societal norms, guide the interpretive process [16], unveiling or concealing potential interactions.

In light of this, augmenting objects with 'superpowers' involves integrating technologies into seemingly analog and inanimate artefacts, enabling them to react to interactions like gestures, vibrations, movement, and radio frequencies, and when necessary, exchange information (IoT). In our approach, these smart attributes of objects are accessible through interfaces that remain dormant and silent until prompted by user interaction.

Therefore, these artefacts should be conceived as interactive mediators between the user and the invisible technology. Our design process involves constant analysis of technological and lifestyle trends and keen observation of everyday life, which are used as drivers for innovation in meaning. This innovation unfolds along three axes: (i) integrating smart technology into objects, (ii) analysing emerging usage patterns to foresee and inspire future interactions, and (iii) benchmarking market dynamics to inspire possible innovation. To fuel these directions and leverage them as design drivers, we adopt an approach that values the significant formal and aesthetic impacts driven by attentiveness to technologies, trends, and market dynamics.

This design approach layers meanings within artefacts, combining perceived and real affordances to enrich user experience. The emphasis is on rethinking traditional aesthetic-functional conventions, particularly in home appliances, through a multi-disciplinary design process that bridges technology, lifestyle trends, and user observation. This mindset continually offers opportunities to challenge and rethink the aesthetic-functional conventions of traditional products, particularly in the home appliance setting. Given that this process is driven by multiple variables subject to research and experimentation (see Chap. 4), it is as innovative as it is critical. The complexity of this approach necessitates a collaborative and interdisciplinary exchange within a multi-backround team, where the designer's role evolves to become an interpreter and aggregator of diverse knowledge [43].

The interaction triggered by such products diverges from the conventional 'what you see is what you get', which constitutes a mainstream approach. Instead, the form often communicates a function that is either a simplified version or completely different from the actual one. This condition situates such artefacts in an area of friction among function, usability, aesthetics, and meaning. Ontologically, design centres around users and their needs [18, 35], requiring a profound understanding of user behaviours, habits, gestures, and even aspirations to challenge them. Addressing these aspects becomes particularly demanding when designing digitally augmented product ecosystems [5, 21, 41], capable of activating unexpected interactions based on their relationship with the user, other objects, or their surroundings, including ecosystems of smart products. In this sense, a defining feature of objects with embedded technology and networked (IoT) is their ability to respond adaptively to various external variables [47]. As a consequence, variables ranging from user inputs to environmental factors, contribute to trigger the transition from passive to active states and engage in a dynamic interaction with their surroundings.

This complex yet intriguing condition is further enriched by designing objects with a minimalist aesthetic rooted in the concepts of materiality and manipulation [28, 48], highlighting the role of aesthetics in triggering meaningful interactions [36]. Here the challenge involves shifting across semantic areas, activated by the augmented nature of the object. Equipping an object with an interface that conceals or disguises undeclared affordances alters its very nature, challenging traditional user perceptions and interactions. Consequently, understanding how users perceive and engage with such objects constitute a primary point in our research.

In a context where a people-oriented design approach prevails, our experimentation deliberately plays with aesthetics, leveraging embedded technology to feature smart objects with multiple and often not immediately perceivable affordances. Hence—and within this specific framework—the leading questions of this chapter are:

- *What happens when an object is consciously designed not to explicitly suggest interaction with it?*

- *How does the semantic reconfigurations of an object impact on its affordances and agency?*

Answering these questions requires exploring the mediating effects of technology on human-object relations [7], thereby embracing a postphenomenological perspective to understand how socio-cultural norms influence the interpretation of interactions [39]. Such a perspective considers technology as an integral and dynamic component of our experience, fundamentally shaping and being shaped by the way we perceive, interact with, and understand the world and ourselves within it. Established interpretations and hermeneutics are influenced by socio-cultural norms, shaping perceptions and interactions and sometimes leading to divergent behavioural outcomes.

By investigating this topic at the crossroads of design, technology, and user interaction, we experiment with the direction of semantic reconfiguration, and focus on embedded technology and its effects on how objects communicate or conceal their functionalities [22, 27]. Exploring semantic reconfiguration involves analysing how embedded technologies in design lead to a shift in the meanings and perceptions associated with objects. This process alters how objects are understood and interacted with, moving beyond their conventional appearances and functions. By integrating digital capabilities discreetly within physical forms, objects begin to communicate and operate in ways not immediately apparent, challenging and expanding our traditional expectations of how they should behave. This reconfiguration not only changes the direct interaction between humans and objects but also impacts the broader understanding of what objects signify within our physical and social environments, thus redefining the dynamics of human-object relationships. As a consequence, this exploration demands a thorough examination of agency and affordance roles during object interactions, especially when traditional affordances are not explicit.

In light of this, this research theme, both controversial and peculiar, is rooted in the concepts of agency and affordance of objects and their interfaces [22, 27], requiring a shift in perspective. Instead of functions being clearly expressed, embedded technology allows for an expansion of potential interactions with seemingly silent objects. This leads to a semantic reconfiguration, where a shift begins with aesthetics and significantly impacts interaction. Reinterpreting the foundational definition of Design as the creative attribution of meaning to things [26] we challenge the ecological approach to the perception of meaning [11, 12, 37], resulting in a misleading design.

5.2 Theoretical Framework

This research has its theoretical underpinnings across four interrelated domains of Interaction Design: variable affordances, embodied interfaces, embedded technology, and expected interaction, thus extending beyond traditional boundaries to include a fourth, integrative aspect. This foundational framework paves the way for examining how objects transform their features by gaining a 'communicative skin', and the implications of information being delegated, embodied, and sometimes concealed within the object. This concept underscores our interest in how information is both manifested and obscured within such objects, thereby enriching the dialogue between user and object through design.

By doing so, the study implicitly broadens its scope and critically looks at the intersection of aesthetics, interaction design, and communication design, questioning and redefining the ecological approach to object design [37]. It investigates how objects, when designed to not explicitly suggest interaction, can challenge and redefine user perceptions of agency and affordances [16]. Hence it delves into the transformative potential of design that does not straightforwardly cue user interaction but rather prompts a reevaluation of established norms.

Spanning from overarching themes to more focused and vertical enquiries, this paragraph explains the domains that delineate the study's scope while outlining its borders. Thus, the following explanation aims at clarifying the research's direction, highlighting its contribution to the broader discourse on interaction design, to a specific area of investigation.

5.2.1 Domain of Reference 1: Variable Affordances

The exploration begins with a fundamental concept. Affordances are defined as the inherent qualities of an object that suggest appropriate actions for manipulation [12]. They are invitations to interact, inherent in every object. High affordance objects clearly communicate their function and usability, while low affordance objects do not. Within the broad landscape of affordances, this study focuses on variable affordances: those that change over time. The primary domain of this research revolves around the dynamic nature of objects that adapt their functions in response to external variables, thereby altering their affordances. Unlike static objects with fixed affordances, those with variable affordances can transition from a passive, unactivated state to an active state, suggesting new interactions with their environment and users. This variable affordance concept challenges traditional notions of object functionality and interaction, highlighting the evolving nature of user engagement with smart objects. This interaction depends on the object's ability to modify its features, thereby initiating a dialogue with its context [29]. Therefore, this study is deeply situated in the discourse on affordances beyond their traditional understanding [4], vetting into the distinction between stable and variable affordances, with the latter being contingent on transient characteristics and closely linked to the specific actions they elicit. The concept of variable affordance is critical in understanding objects that possess the ability to alter their functionalities, akin to a communicative skin, thereby enabling varied forms of agency [45, 46]. This kind of interaction is facilitated by technological enablers like sensors and IoT connectivity, which allow objects to perceive and react to user behaviours and environmental changes. Agency refers to the capacity of both users and objects to initiate and control actions, pointing out the active role of objects in the interaction process—beyond being a response to user inputs. In this domain, agency is impacted by variable affordances, where objects are not just passive recipients of user actions but are capable of influencing and guiding user behaviour through their adaptive features and functionalities.

While variable affordances offer dynamic interaction opportunities, they also pose challenges to user agency through potential ambiguities in affordance interpretation. Users may face difficulties in predicting or interpreting object behaviour as affordances change, leading to a possible mismatch between user expectations and actual object functionality. This uncertainty can hinder the user's ability to grasp the object's function, requiring additional effort to understand and adapt to its evolving functionalities. In light of this, the tendency is to designerly address this challenge by embedding clear feedback mechanisms, from visual to auditory and haptic signals, to guide users through the interaction, ensuring that the evolving functionalities of objects enhance rather than complicate the user experience.

5.2.2 Domain of Reference 2: Embodied Interfaces

The concept of embodied interfaces emerged in the 1990s, fundamentally redefining the interaction between users and digital devices [9]. Initially conceptualised from the idea of invisible interfaces, they rely on more intuitive forms of interaction, often through physical manipulation to interact with an artefact, for instance by tilting, translating, and rotating it [9]. Embodied user interfaces leverage manipulation as a means of interaction, as opposed to tapping on a screen or pressing buttons. A subset of such interfaces is Tangible User Interfaces, where physical objects are used as tangible representations of digital information [19], functioning as controllers linked to virtual representations. This approach bridges the gap between the physical and digital worlds, making the interaction more tangible and less abstract.

The implementation of these interfaces generally involves sensor technologies for detecting and interpreting physical actions, empowering the object to understand and respond to manipulations and perform functions without the need for traditional interfaces.

In this framework, information is not merely displayed but is embodied within the object, often remaining concealed until activated by the user. This paradigm shift places a significant emphasis on the user experience, where interaction is not just about functionality but also about the engagement with the physicality of the object. Embodied interfaces blur the lines between the digital and physical realms, creating a more immersive and intuitive user experience.

The discussion focuses on the implications of such interfaces, particularly how information is delegated, embodied, and sometimes masked within the object [8]. The notion of an interface being a part of the object's body itself, rather than a separate element, challenges traditional interface design and prompts a reconsideration of how users perceive and interact with technology. One challenge of embodied interfaces is ensuring that physical actions are interpreted accurately by the technology, avoiding misinterpretations that could frustrate users. Among the related challenges, the risk that users with different physical abilities may find these interfaces less accessible, potentially limiting their agency. While embodied interfaces aim to enhance agency by making interactions more natural and intuitive, they also require special reflection on inclusivity and accessibility to ensure that all users can use them. Also cultural interpretation plays a role in how such interfaces are perceived and interacted with, introducing another layer of complexity to their design and use. Different cultures have varying norms and expectations regarding physical interaction, space, and gestures, which can significantly influence how an embodied interface is received and understood. Cultural variability can challenge the universality and accessibility of embodied interfaces, potentially limiting user agency for individuals from diverse cultural backgrounds. Especially in a market progressively borderless and with more and more intercultural populations, carefully considering cultural nuances becomes key.

5.2.3 Domain of Reference 3: Embedded Technology

This domain explores the transformative potential of embedded technology in redefining object interaction and manipulation. Embedded technology refers to the integration of technological components, such as sensors (both individual and networked), along with software, that are embedded within objects. Such technology provides objects with the capability to sense and respond to external stimuli, including environmental changes.

The relevance of embedded technology lies in its capacity to extend the levels of interaction with seemingly inert objects. Through the incorporation of sensors, which are initially invisible to the user, these objects acquire new dimensions of interaction. In the IoT context, these embedded systems play a critical role in connecting everyday objects to the internet, allowing them to send and receive data [33]. This capability transforms these objects into dynamic agents of interaction, capable of responding in novel and often unexpected ways to user actions and environmental contexts [6]. The inclusion of IoT principles in this domain broadens the scope of interaction beyond direct user inputs, enabling objects to interact intelligently with each other and with their environment.

Embedded technology thus serves as a gateway to a world where objects are not merely passive but are active participants in the user experience. This interaction is not limited to direct user inputs but extends to the objects' ability to perceive and respond to their surroundings, creating a more responsive and adaptive environment where user experience is enhanced through seamless technology integration [38]. The exploration in this domain is crucial for understanding how technology can be seamlessly integrated into everyday objects, enhancing their functionality and user interaction without compromising their aesthetic appeal or usability.

Additionally, the convergence of embedded technology, as wired and wireless sensors, with advanced networked systems and data analytics can further augment the smartness and intelligence of objects, equipping them with the power to anticipate needs and adapt their behaviour proactively. This predictive capability, enabled by machine learning algorithms and big data analytics, adds a new layer to the user experience, offering personalised and context-aware interactions [1].

The incorporation of embedded technology within objects fundamentally alters affordances by endowing objects with the capacity to perceive, process, and react autonomously to environmental stimuli and user inputs. This technological enrichment challenges traditional notions of static affordances by introducing dynamic, context-sensitive functionalities that can adapt in real-time to suit specific needs or conditions. While this enhances objects' utility and responsiveness, thereby potentially increasing user agency by providing more tailored and intelligent interactions, it also introduces complexities in understanding and predicting object behaviour of technology within objects and introduces the challenge of ensuring that these objects can accurately sense and respond to user actions and environmental conditions. For instance, misinterpretations by sensors or algorithms can lead to unexpected behaviours, undermining user agency by providing responses that do not align with user intentions or needs. The challenge for user agency arises when the sophistication of embedded systems makes it difficult for users to anticipate how objects will respond in different situations. This unpredictability can lead to a sense of loss of control or confusion about how to achieve desired outcomes, as users may not always understand the logic behind the object's actions.

5.2.4 Domain of Reference 4: Expected Interaction

This domain examines the activation of user behaviours and interactions through the recognition of familiar patterns and cues. It delves into how the inherent expectations and norms, deeply embedded in societal and individual contexts, guide the interaction between users and objects. Drawing upon the seminal works of [10, 13], this discourse extends from social science theories to elucidate how users approach and engage with technology.

Garfinkel [10] explored 'scripts' in social interactions. Although the discourse is situated in social science, it can be extended to how certain features activate sequences of actions because they are recognised. In the context of this discourse, his concept can be applied to how users interact with objects. Garfinkel's scripts refers to the set of expectations and norms that people rely on in everyday situations. When it comes to interacting with objects, especially those with embedded technology, users often have preconceived scripts or expectations about how these objects should function based on their past experiences and societal norms. Objects with hidden or variable interfaces challenge these scripts, as they may not conform to the usual cues and patterns that users expect. This leads to a re-evaluation of how users approach and understand these objects, potentially leading to new forms of interaction and engagement. Goffman [13], on the other hand, focused on the patterns of social interaction and the presentation of self in everyday life. His work can be applied to understand how objects fit into the broader patterns of user behaviour and social interaction. In the context of design, Goffman's perspective can be applied to analyse how objects with unconventional interfaces integrate into or disrupt established social and behavioural patterns. For instance, an object that does not immediately reveal its function may alter the typical pattern of interaction, requiring the user to engage in a process of discovery and learning. This can lead to a shift in the user's perception and behaviour, affecting the social dynamics around the use of the object.

The primary challenge is the potential dissonance between user expectations and the actual interaction experience. Namely, the disruption of traditional interaction scripts, which can confuse users and challenge their agency. Such dissonance can momentarily disrupt the user's sense of competence and control, impacting their agency negatively. Indeed, when objects do not conform to established norms and expectations, users go through experiences which can be both disorienting or enlightening. This disruption requires users to question and learn new interaction patterns, thereby expanding their competency and agency in the long term. However, the initial confusion and potential frustration represent significant hurdles to user adaptation and acceptance, requiring thoughtful design to bridge the gap between novel and familiar interaction paradigms. By confronting and overcoming these challenges, users can develop a broader understanding of interaction possibilities, thereby enriching their interaction repertoire. In terms of affordances, expected interaction invites a reevaluation of how objects communicate their potential for use.

Table 5.1 summarises the theoretical framework so far discussed, presenting a concise overview of the three domains of reference, their core concepts, the focus of the study within each domain, the technological aspects involved. It outlines the main concepts central to each domain, focusing on the dynamic interaction between objects and users mediated through technological advancements, and then clarifies how each domain specifically affects affordances and agency. Key aspects include the adaptability of object affordances, the intuitive physicality of user interfaces, and the transformative potential of embedded technologies within objects. In light of this, implications on affordances discusses the changing nature and perception of affordances in each domain, while implications on agency focuses on how the agency of both users and objects is affected, highlighting the evolving dynamic between users and technology.

Positioned at the intersection of these four domains of reference, smart objects equipped with hidden and variable interfaces introduce distinct challenges for design and user experience, necessitating thorough reflection. In digitally augmented objects, the aesthetic dimension is enhanced, and interfaces play an essential role in moulding and directing user understanding. Enabling actions through *operative* [42] and *fatic* [20] functions, interfaces play a pivotal role in informing and orienting user interaction.

Relying on the post-phenomenological lens introduced above [7], technology-augmented objects present a challenge to established design conventions, especially regarding socio-cultural and normative considerations [15]. The strategy of embedding information within the object, effectively hiding it until user interaction occurs, links affordances and behavioural outcomes to a silent layer. As such parts of the affordances are not revealed by design. This approach raised critical questions about

Table 5.1 Key concepts, focuses, technological underpinnings within each domain, and then a focus on implications the domain implies in terms of affordances and agency

Domain	Main concept	Key focus	Technological aspects	Implications on affordances	Implications on agency
Variable affordances	The dynamic nature of objects that adapt their functions in response to external variables, altering their affordances	Exploring how objects can transition from a passive to an active state, suggesting new interactions	Technology enabling objects to change affordances	Affordances become fluid, adapting to context and user interaction	Users gain dynamic control over object functions, enhancing interactivity
Embodied interfaces	Interfaces that rely on intuitive forms of interaction through physical manipulation, bridging the physical and digital worlds	Enhancing user experience by making interactions more tangible and intuitive	Sensor technologies for detecting and interpreting physical actions	Makes affordances more intuitive and physically engaging	Empowers users through more natural and intuitive interactions
Embedded technology	The integration of technology within objects, enabling them to sense and respond to external stimuli	Transforming objects into dynamic agents of interaction through embedded systems and IoT	Sensors, IoT connectivity, machine learning algorithms, and big data analytics for predictive capabilities	Enhances objects' ability to present context-specific affordances	Objects become proactive participants, influencing user behaviour and choices
Expected interaction	Activation of user behaviours and interactions through recognition of familiar patterns and cues, guided by societal and individual expectations and norms	Understanding how users' preconceived scripts about object functionality influence their interactions	Not directly specified, but involves the design that influences user recognition and interaction patterns	Challenges and reshapes user expectations about object affordances	Requires users to adapt, potentially expanding their capacity for agency in novel interactions

the impact of aesthetic interventions on object interpretation, and required necessary reflection on the implications of redesigning objects that alter their perceived meanings, intentionally inducing cognitive dissonance that influences both the user interface (UI) and the user experience (UX).

5.3 When Design is Misleading

Building upon this framework, the chapter conducts a focused exploration into the implications of redefining meanings through design. It follows a design-driven innovation approach that prioritises design experimentation inspired by the potential of emerging technologies [44]. This approach diverges from traditional user-need-driven design. This departure is nurtured by the prioritisation of technological potential and aesthetic possibilities over immediate user requirements. In traditional design approaches, the starting point typically involves identifying and addressing explicit user needs, ensuring that the design solutions directly respond to these identified problems or desires. Conversely, the design-driven innovation discussed in this study begins with a focus on the capabilities and opportunities presented by emerging technologies, exploring how these can be harnessed to create novel interactions and gratifying experiences. This approach places a significant emphasis on the creative and experimental aspects of design, allowing for the development of products that may not have been directly requested by users but that offer new forms of engagement and interaction. The exploration of dynamic and hidden interfaces, in particular, introduces a layer of interaction that is not immediately apparent, necessitating a rethinking of how users discover and engage with functionalities. This approach can lead to innovative products that redefine user experiences and interactions, potentially creating new needs or transforming existing ones.

This analysis is articulated through the examination of several cases, each exemplified at the convergence of the study's four foundational domains: (1) Variable affordances, (2) Embodied interfaces, (3) Embedded technology, and (4) Expected interaction. These instances demonstrate varying degrees of *misleading-ness*, namely ambiguity inherent in objects that do not explicitly communicate their affordances. The examples range from designs concealing unanticipated functionalities to those undergoing semantic reshaping to integrate aesthetics with functionality.

'Shape' is understood beyond its physical dimension to encompass design elements that significantly influence how an object communicates its function and guides user interaction. This criterion assesses how the design's visible features signal to users what the object does and how it should be used. The variability in these design elements' clarity can either elucidate or obscure the object's affordances, directly impacting how users perceive and engage with the object. Thus, 'shape' is identified as a criterion for discerning the cases, not merely in terms of physical form but as an aspect of design that includes elements varying in clarity and thus affecting the communication of an object's affordances and its operational guidance. As shown in the figure, 'shape' is used to express how effectively a design communicates its intended use and functionality, highlighting the intricate relationship between form, function, and user experience.

The analysis focuses on how these objects invite interaction, challenge design conventions, and initiate a dialogue with users, prompting new forms of engagement and interpretation. In the analysis, the selected examples are arranged on a Cartesian axis to illustrate the relationship between the design's form and its functional intent, as well as the impact on communicative effectiveness. The horizontal axis measures the degree to which the physical form of an object aligns with its intended function: how closely the design's appearance (form) suggests its purpose (function), highlighting the relationship between how an object looks and what it does. The vertical axis illustrates the extent to which the object's design and aesthetics impact its capacity to communicate its functionalities and affordances: their ability to convey to users how they should be interacted with. Figure 5.1 positions some examples that will be discussed in Sect. 5.4 along two axes thereby. This representation is meant to facilitate the understanding of how different designs with varied balancing of aesthetic and functional aspects impacts on the communication of their affordances. Objects placed towards the right on the axis are those whose form closely mirrors their function, making their use intuitive. Those higher on the vertical axis possess a greater capacity to communicate their functions, thereby simplifying user interaction.

This analysis further explores these examples in relation to the three distinct directions identified to describe the process of designing smart objects with concealed and variable interfaces. In Sect. 5.4, advancing from this representation, it is examined how smart objects play with the delicate equilibrium between innovative design and user understanding. It sheds light on the challenges these objects pose to conventional design norms, exploring how they redefine user interaction by requiring users to

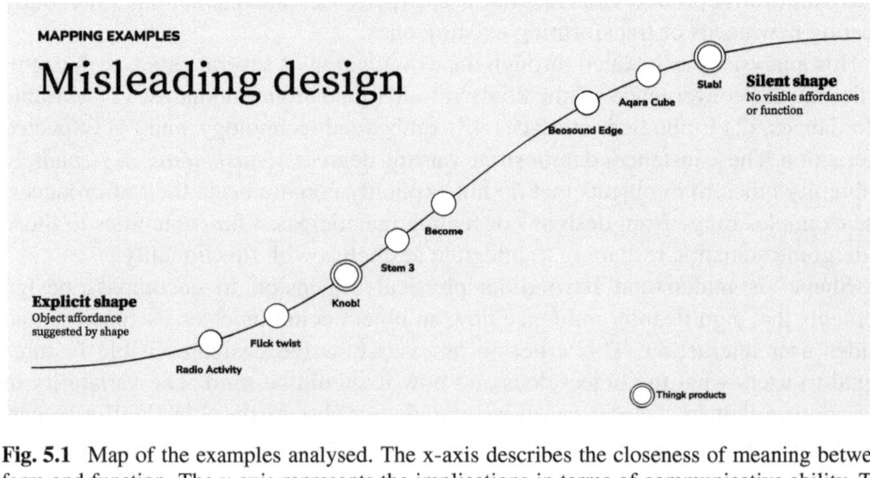

Fig. 5.1 Map of the examples analysed. The x-axis describes the closeness of meaning between form and function. The y-axis represents the implications in terms of communicative ability. The second stroke identified Thingk's products

engage in a deeper level of interpretation, discovery, and engagement, thus fostering new and potentially more enriching forms of interaction.

This selection of examples blends well-known and proprietary cases, and serves to critically analyse the design implications of objects that, at first glance, appear non-functional. These case studies are instrumental in unfolding how challenging conventional design norms and established behavioural patterns, can significantly alter user interaction. The analysis delves into both the benefits and drawbacks of designing objects that defy traditional expectations and norms. By doing so, it sheds light on the complexities of interaction design where the usual cues for functionality are absent or misleading, thus providing insights into the broader implications for user experience. This exploration is grounded in an ongoing discourse on how social behaviours and interactions are activated by recognition of partners [10, 13].

5.4 Exploring Design Implications

When a design intervention focuses on the aesthetic aspects of objects from a semiotic standpoint, it introduces significant implications.

The capacity of an object to signify a function through its design, serving as a meaningful attribute, is transformed and reconfigured by design. The relationship between form and function becomes less overt and direct. Objects are intentionally crafted to engage in technological mediations that become apparent during user interaction. As the interface—acting as a sign—is recognized, objects unveil their functionalities, facilitating their interpretation and engagement, thereby soliciting user agency. In the following, we critically examine the consequences of incorporating technology into objects that appear inert, initiating a discussion on technological enhancement [45, 46]. It acknowledges that affordances invite specific behaviours, thereby interfering with the interpretative process. Objects are designed to challenge users' understanding of their function and operation. Nevertheless, once the interactive capabilities are exposed, such technological mediation can foster satisfaction and even contribute to the development of new literacy.

Figure 5.2 extends the analysis initiated in the precedent paragraph by integrating the relationship with three distinct directions we have delineated from the analysis of the design practices, which characterise the approach to designing smart objects with hidden and variable interfaces. These directions—Emphasising shape-function, Challenging design conventions, and Semantic reconfiguration. All together they underscore diverse strategies employed in the design of smart objects with hidden interfaces, and will be elaborated upon in the following discussion.

The following analysis delves into three primary design directions that emerge in the creation of smart objects with hidden and variable interfaces, which reveal themselves through user interaction, thus shaping the perception of object affordances. Employing a range of examples, this section examines the design characteristics and their impact on eliciting user interpretation, thereby influencing the perception of the object. This exploration transitions from the misleading attributes of these objects

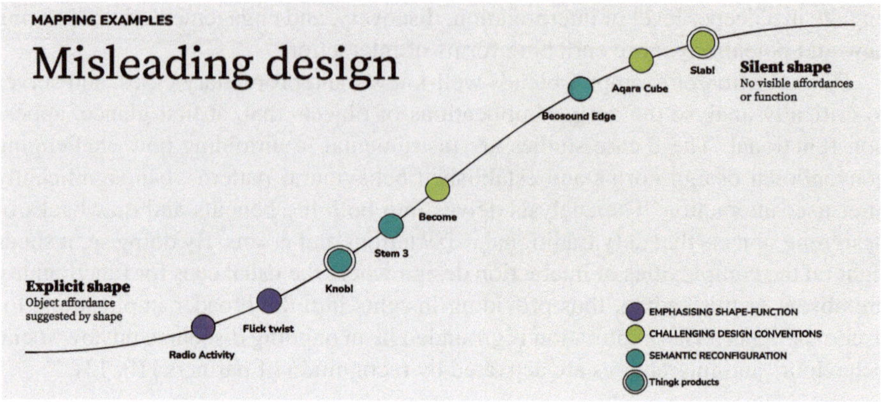

Fig. 5.2 Map of the examples analysed associating them to the three diverse strategies employed in the design of smart objects with hidden interfaces

to their design implications. We consider it noteworthy to point out that these three directions are not mutually exclusive but can coexist within the same object.

5.4.1 *Emphasising Shape-Function*

The first direction revolves around the concepts of verisimilitude and emphasis intended as the strategic accentuation of design elements.

Flick Twist by Flic (2024; flic.io/twist), unveiled at CES 2024, exemplifies an advanced smart home controller. It integrates four distinct interactions—push, double push, twist, and push and twist—to facilitate comprehensive customisation across a spectrum of domestic devices including lighting, audio systems, window blinds, among others. Despite its minimalist aesthetic, which does not directly suggest any specific function beyond that of a general controller, *Flick Twist* embodies the archetype of a volume control. Through the sole use of rotation and pressure as intuitive and conventional means of interaction, it allows users to execute a variety of actions and digital commands through diverse action combinations. While it presents itself in the form of a traditional dimmer or controller, it subtly implies its multifunctionality without explicitly disclosing the full range of its capabilities.

A second and more challenging and significant object for this first direction is *Radio Activity* by Gemma Roper (2015; gemmaroper.com/Radio-Activity). It is an internet-connected radio that synchronises with Spotify to select music based on beats per minute, employing a control mechanism that emulates a metronome for adjusting volume and tempo. This device's minimalist design deliberately mimics the appearance of traditional metronome for tempo, thereby leveraging verisimilitude to suggest its functional capabilities without overt exposition. The overtly reduced aesthetic is a key aspect and it plays on verisimilitude with forms traditionally associated with the two functions.

These instances illustrate a deliberate design strategy where the iconic function is not merely represented but is imbued with deeper significance, effectively becoming the interface itself. This approach encourages users to engage in a deductive process, piecing together operational cues through the interpretation of a semantic void intentionally left by the designer. Such design decisions are deeply rooted in socio-cultural conventions that serve as scripts and reference frameworks guiding our interaction with the world. This approach not only acknowledges established patterns of usage and the collective reservoir of knowledge, but exploits them to convey additional meanings in a subtle manner. As a consequence, this approach frames design as a deliberate act of sense-making within the confines of culturally defined definitions of reality [10, 13].

5.4.2 Challenging Design Conventions

This direction delves into the juxtaposition of design elements that conceal advanced functionalities beneath ostensibly simplistic exteriors, challenging established norms of design. The example *Become* by Rlon (2018; rlon.com/become) serves as a paradigmatic instance, consisting of a desk lamp that is activated through the placement and movement of a metal sphere upon a black monolith, engaging with varying magnetic fields to illuminate. This instance markedly shows the widening breach between form and operational function, where the act of positioning the sphere near a designated area functions as the mechanism of activation.

An additional illustration is provided by the *Aqara Cube T1 Pro* (2022; aqara.com/en/product/cube-t1-pro), a six-sided wireless controller that signifies a leap forward in the field of home automation. This controller simplifies the complexity of home device management by adopting intuitive gestures. The orientation of the cube's faces becomes the trigger to enact commands as active gestures, showing high versatility and customizability. *Aqara Cube T1 Pro* is designed for an advanced user experience, positioning it not merely as a control device but as a harbinger of more interconnected and intelligent home environments.

Our experimentation in this direction is exemplified by *Slab!* (Thingk, 2017). It adopts the guise of a commonplace wooden block while covertly functioning as both a kitchen scale and a digital timer. This object conceals its multifaceted functionality within a form familiar to the kitchen setting. The user's manipulation of the object's orientation—horizontal for scale functions and vertical for timing—reveals its dual capabilities, with tactile interaction prompting the display's appearance (Fig. 5.3). By touching the surface, the display emerges and, by changing its orientation, the same display becomes a timer for controlling the cooking time. Such design effectively portrays a seemingly silent object that conceals its complexity, demonstrating the potential for smart technology to subvert traditional design expectations.

Fig. 5.3 *Slab!* functioning as a kitchen scale and then timer when re-oriented

Being overall aligned to the aforementioned post-phenomenological perspective [7], these examples underscore the capacity of technologically augmented objects to challenge and even subvert foundational design principles, including ergonomics and the socio-cultural codification and normative of object use. This analytical perspective reveals how contemporary design strategies, by embedding advanced functionalities within deceptively simple interfaces, can embed cognitive dissonance, thus recalibrating conventions (patterns and scripts) and expectations rooted in users [26].

5.4.3 Semantic Reconfiguration

The last direction concerns the semantic reconfiguration of objects, wherein their design intricately blends aesthetics and functionality. A prominent and well-known case to discuss this concept is the *Beosound Edge* (Bang and Olufsen, 2018, https://www.bang-olufsen.com/en/gb/speakers/beosound-edge).

The speaker itself serves as an interactive controller requiring rotational movements to manage the music and adjust its settings. This integration introduces an additional and complementary interface on the speaker's surface, enriching the user experience by making the operational functions overt during interaction. The design embodies the essential characteristics of its functionality, integrating elements typically associated with control and interaction into the physical form of the speaker. The *Beosound Edge* speaker represents a striking instance of semantic reconfiguration in design, challenging conventional user interactions with audio devices. Traditionally, speakers are passive objects, manipulated through external controls or interfaces such as buttons, dials, or digital touchscreens. However, this speaker reimagines this interaction paradigm by incorporating the control mechanism directly into the physical form of the speaker itself. Rather than relying on detached controls, it invites users to physically engage with the speaker by pushing it to manage music playback. This action—rotating the speaker along its axis to adjust volume or change tracks—is a significant departure from standard interaction models. It transforms the speaker from a mere output device into an interactive object that responds to user input through physical movement. This design choice not only enhances the aesthetic appeal of the speaker but also embeds a layer of functionality that is intuitively discovered through interaction.

The act of pushing the speaker to control music leverages the user's natural inclination to interact with objects in their environment, yet it subverts expectations by assigning a novel function to this familiar gesture. This reassignment of function to form necessitates a reevaluation of how objects communicate their use, inviting users to explore and engage with the speaker in a manner that is both exploratory and rewarding. In doing so, the *Beosound Edge* exemplifies semantic reconfiguration by merging aesthetic elegance with functional innovation, creating a seamless integration of design and interaction that challenges and expands the traditional boundaries of user experience with technological devices.

Another instance of innovative design is the *Stem 3* developed by Kano Computing (2022; stemplayer.com/stemprojector), the company that created the *Stem Player*. This portable device comes with a selection of movies, with the option to upload your own to its 128 GB of built-in memory or to stream them via AirPlay or Chromecast. The projector enables content manipulation through touch controls offering haptic feedback. It can be controlled via touch buttons, allowing for the modulation, composition, and remixing of all produced content through specific gestures and various light interfaces. Users can use the *Stem 3* to modify the multimedia content in real-time, altering visual elements like colour palettes, adjusting playback speed, or changing the screen shape, thus facilitating an interactive and creative engagement with the media. Thanks to various lighting configurations that shine through its communicative skin, via multicoloured LEDs, this object allows for touching and using multiple functions in ever-changing ways, activating the numerous features included in the user experience it can offer.

It serves as an example of semantic reconfiguration by redefining the user interaction model for multimedia content projection and manipulation. Unlike traditional projectors that typically rely on external input devices (like remote controls or connected computers) for operation and content management, the *Stem 3* integrates interaction directly into the device itself through touch-sensitive controls and gesture recognition. Its semantic reconfiguration implies a shift from the conventional, button-based interaction paradigm towards a more intuitive, gesture-based control scheme. Users can modulate, compose, and remix audio-visual content through specific gestures and interactions with various light interfaces on the projector itself. This means that actions such as swiping over the projector's surface, adjusting its orientation, or even moving one's hands in proximity to the device can alter contents. This design choice transforms the projector from a simple display device into an interactive tool for creative expression, allowing users to engage with their media in dynamic and innovative ways. The inclusion of haptic feedback further enhances the tactile experience, providing immediate, intuitive responses to user inputs. By embedding these advanced functionalities within a seemingly straightforward device, the *Stem 3* challenges users' expectations about how projectors should operate. This reimagined approach to controlling and manipulating multimedia content redefines the relationship between form and function, encouraging users to discover and leverage new modes of interaction that extend beyond the traditional confines of projector use.

This direction critiques established design paradigms by focusing on objects whose interactive mechanisms are not immediately apparent, challenging the traditional conceptions of agency and affordance [11, 12]. These examples underscore a significant paradigm shift, where the interaction mechanism is seamlessly integrated into the object's form, representing a semantic reconfiguration that not only questions [26] foundational definition of Design but also contests the ecological approach to meaning interpretation advocated by [12, 37], resulting in a misleading design that deliberately obscure their mode of interaction.

Advancing this discourse, considerations were given to the potential of embedded interfaces that are both variable and responsive to environmental contexts, specifically asking ourselves: What if the embedded interfaces became variable and situated? This speculation was further developed through participation in the EU-funded Decochrome project (cordis.europa.eu/project/id/760973), leading to the conceptualisation of interfaces that adapt to user interactions and environmental changes. We imagined a model of interfaces featuring high adaptability and context-awareness in its design. This innovative approach is exemplified in *Knob!* (Thingk, 2019; Fig. 5.4) a device that employs a context-sensitive interface that adjusts its functionality based on its placement and orientation but also to the surrounding environment, its variables and parameters, and eventual smart object [32]. Once paired with a technological bridge, such as a smart home hub or application, that connects it with other smart devices in the environment, the object can serve as a volume control when positioned on a table, a thermostat controller when affixed vertically to a wall, or an ambient light dimmer when placed on a bedside table. The concept inherently plays with design conventions that lead to interpreting objects according to their position. As such, it represents a significant departure from traditional, static interfaces by incorporating dynamic responsiveness not just to direct user inputs but also to the broader context of its environment, including the presence of other smart devices. The core idea revolves around creating interfaces that are inherently flexible and capable of altering their functionality based on their physical positioning and the specific requirements of the interaction scenario. This approach leverages design conventions that intuitively signal how an object should be used based on its location and orientation, encouraging users to engage with objects in a manner that is both exploratory and intuitive. In practical terms, this concept was materialised through the design of a cylindrical object, a form chosen for its simplicity and ease of manipulation. This cylinder embodies the notion of a variable, situated interface, a term that fully embeds its ability to modify its function in response to its situational context.

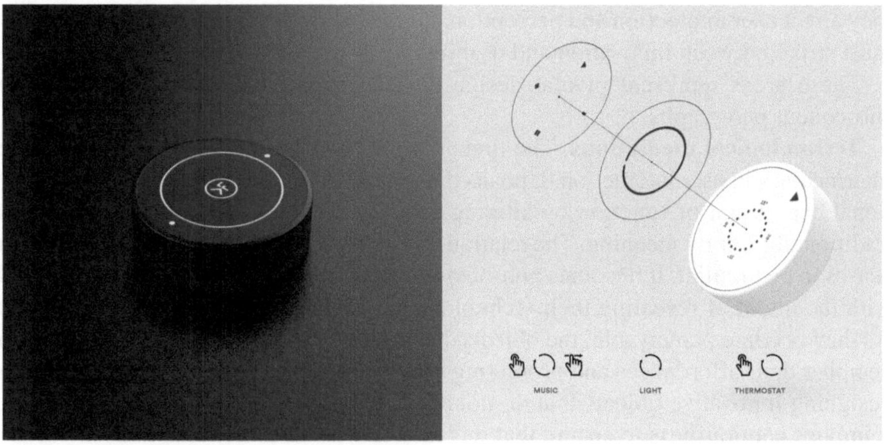

Fig. 5.4 *Knob!* and its variable interfaces, which surface and change according to the positioning in the environment and the interaction with other objects

This design exemplifies semantic reconfiguration by challenging and expanding the traditional roles assigned to objects within interactive systems. This reconfiguration is built on a form of the object capable of adapting to multiple functions. The attribution and perception of the object's meaning depends on variable elements. The paradigm shift starts with aesthetics, triggered by spatial position, orientation, or dialogue with other smart objects, and ends impacting interaction. By assigning multiple, context-dependent functionalities to a single form, the *Knob!* goes beyond conventional interface paradigms. It not only responds to direct user actions but also to its placement within an environment, thereby achieving a higher degree of integration and interaction with the user's everyday life. This approach significantly enriches the user experience by providing a seamless, intuitive means of controlling various aspects of the ambient environment, demonstrating a sophisticated application of design thinking that bridges physical interaction and digital technology.

5.5 Design Issues: Discussing Challenges

In scenarios where objects are intentionally designed without clear indications for interaction, they present unique challenges to users. Drawing from the preceding discussion, four critical and expansive areas of reflection emerge in the design of such misleading objects.

Table 5.2 captures the essence of each design issue identified, defining the concept and exploring its implications for how users perceive and interact with objects. It reflects on the dynamic interplay between design, technology, and user experience, underscoring the potential of misleading objects in reshaping user interaction while pointing out inherent challenges in implementing these design philosophies, and how they affect user interaction and perception. It highlights the careful balance designers must strike between innovation and usability.

These areas represent pivotal design considerations and are examined in the subsequent paragraphs.

Technological mediations. The first area concerns the role of technology as an intermediary in user interaction. It posits that technology can reconfigure the conventional expression of function by altering established practices and designs, which traditionally signify meaning. The relationship between an object's form and function ceases to be familiar. It becomes non-obvious and indirect. These objects are crafted with the intent of revealing their technological capabilities during user interaction. As they become perceivable, the objects disclose their true nature, enabling users to decipher their affordances and prompting user agency. The primary challenge lies in designing innovative objects that do not alienate users. From a design perspective, it implies cautiousness to ensure that this indirectness does not lead to user frustration or confusion. Ensuring clarity in how technological capabilities are gradually revealed is key to maintaining engagement without overwhelming the user. A critical consideration is the balance between innovation and usability. Objects designed under the paradigm of technological mediations should foster a sense of exploration and engagement, encouraging users to uncover functionalities not immediately apparent while avoiding to fall into unpleasant or obscure interactions.

Physical storytelling. This concept explores the augmentation of digital interfaces with analog experiences, striving for a novel tactile dimension. By blending the tangible aspects of analog design with the intangible qualities of digital technology, it

Table 5.2 Design issues, associated challenges, and effect on users

Design issue	Concept	Challenge	User impact
Technological mediations	Technology as an intermediary, making form-function relationships indirect	Balancing technological innovation with intuitive user interfaces	Requires users to explore and engage deeply to uncover functionality, enhancing discovery
Physical storytelling	Blends digital and analog for a tactile experience, merging simplicity with tech	Designing interfaces that are both intuitive and capable of conveying complex digital functions in a physical form	Facilitates intuitive use but may initially confuse users expecting traditional interactions
Semplexity	Aesthetic minimalism with hidden technological complexity	Achieving a minimalist design without sacrificing functionality or user accessibility	Delights with unexpected features but risks overwhelming users with hidden complexities
New literacy	Misleading designs that introduce new interaction modalities	Creating objects that challenge but do not frustrate user expectations and capabilities	Enhances engagement through discovery but requires adaptation to new forms of interaction,

aims to guide user expectations through familiar interactions. This approach requires designers to thoughtfully integrate the tangible elements of analog design with the abstract qualities of digital technology. The challenge concerns guiding user expectations and interactions through familiar yet innovative means, thereby embedding a deeper understanding and appreciation of the technology within the user experience. Articulating digital complexities into intuitive physical forms calls for detailed examination of material choices, form, and the interaction design to ensure that physical attributes do not merely encase digital content but actively contribute to and enhance the user's experience. Sensory cues—such as texture, weight, and movement—are critical in crafting an interactive narrative that makes the digital aspects of technology both accessible and engaging. Linking innovation to familiarity constitutes a pivotal aspect of physical storytelling. By introducing users to novel interactive paradigms while grounding these experiences in recognisable physical interactions favours reaching a new equilibrium that engages users with technology lessening the feeling of being overwhelmed by its complexity. Specific attention should be drawn on how physical attributes can invite interaction and use, unveiling the digital dimensions in a manner that feels both satisfying and intuitive.

Semplexity. The third area concerns the harmonious coexistence of simplicity and complexity in objects. This duality goes beyond aesthetic minimalism masking advanced functionalities; it concerns a sophisticated design intervention that requires deep consideration of how form communicates. Designerly speaking, it requires to move along the thin line between minimalism and functionality, ensuring that the semplexity does not alienate users with its hidden complexities. The goal is to delight and surprise users with the object's capabilities without causing frustration or requiring extensive effort to uncover these features. The challenge extends beyond the visual or tactile experience. Extending to the entire user journey, it implies a seamless transition from the simple to the complex, guiding users through a discovery process that feels natural and engaging. This requires not just technical skill but a deep understanding of user psychology and interaction patterns. This discourse also opens a reflection on narrative: traditionally, each object tells a story through its use, gradually revealing its capabilities and encouraging further exploration. Although recognizing that this narrative should be carefully crafted to ensure that users are neither overwhelmed at the outset nor become complacent with superficial functionalities, the specific design intervention related to hidden interfaces counter this storytelling principle, leaving to users the task to advance in a progressive, and not invited by explicit affordances, realisation of the object's functionalities. In this context, designers should pay close attention to feedback mechanisms within the object that reward curiosity and exploration.

New literacy. The fourth issue deals with the new modalities of use that misleading objects require and trigger. The starting point is that affordances have sociocultural and normative aspects. Rooted and established affordances imply the presence of precise frames of references and scripts which are activated when needed. Interfering with the hermeneutic process, these objects are designerly conceived to deceive users about their function and functioning. Therefore, when technology mediates the perception of affordances and behavioural outcomes [7], it can open up new and surprising modalities of interactions, requiring existing frames and scripts to be updated. Once the interaction is unveiled, such a mediating power of technology can also generate satisfaction and gratification, contributing to generating new literacy. From a designerly standpoint, the potential for such designs to challenge established norms and user expectations requires careful consideration of how these new modalities can be introduced without leading to user alienation. Ensuring that new affordances are discoverable and learnable without excessive trial and error is crucial. A direction is that of fostering a sense of discovery and satisfaction in learning to use the object, rather than confusion or frustration.

Drawing from extensive empirical research and design experimentation, the aim of this chapter was to provide a thought-provoking examination of how objects, through their design and embedded technologies, can engage users in new and unexpected ways. The focus was on how these design choices influence the user's perception, interaction, and understanding of objects in their environment, proposing a novel perspective on the future of interaction and interface design. The main intent was then to go through the design challenges behind these objects, with the aim to unpack the multitude of implications that playing at the crossroad of smart objects with minimal aesthetics raise.

In light of this, it becomes evident how, by designing 'misleading', namely crafting objects that diverge from anticipatory norms, designers not only challenges established patterns and contest entrenched design conventions but also foster the emergence of novel technological and interaction literacies, enhancing user engagement and satisfaction [15].

References

1. L. Atzori, A. Iera, G. Morabito, The internet of things: a survey. Comput. Netw. **54**(15), 2787–2805 (2010). https://doi.org/10.1016/j.comnet.2010.05.010
2. S. Boess, H. Kanis, Meaning in product use: a design perspective E, in *Product Experience* ed. by H.N.J. Schifferstein, P. Hekkert (Elsevier, San Diego, 2008), pp. 305–332. https://doi.org/10.1016/B978-008045089-6.50015-0
3. G. Bonsiepe, *Dall'oggetto All'interfaccia. Mutazioni del Design*, ed. by F. Costa (Trans.). (Feltrinelli, Milano, Italy, 1995)
4. A.M. Borghi, L. Riggio, Sentence comprehension and simulation of object temporary, canonical and stable affordances. Brain Res. **1253**, 117–128 (2009). https://doi.org/10.1016/j.brainres.2008.11.064
5. G. Brugnoli, Designing smart experiences. The smart and invisible future of interactions and services. Mediu. (J. Electron.) (2015). https://medium.com/@lowresolution/designing-smart-experiences-a6e675b414ec
6. H. Chaouchi, *The Internet of Things: Connecting Objects to the Web* (Wiley, New York, NY, 2013)
7. N.A. Döbler, C. Bartnik, Normative affordances through technology: a postphenomenological perspective, in *New Trends in Disruptive Technologies, Tech Ethics and Artificial Intelligence*, ed. by J.F. de Paz Santana, D.H. de la Iglesia, A.J. López Rivero (Springer International, Cham, Switzerland, 2022), pp. 145–156. https://doi.org/10.1007/978-3-030-87687-6_15
8. P. Dourish, *Where the Action is: the Foundations of Embodied Interaction* (The MIT Press, Cambridge, MA, 2004)

9. K.P. Fishkin, T.P. Moran, B.L. Harrison, Embodied user interfaces: towards invisible user interfaces, in *Engineering for Human-Computer Interaction: IFIP TC2/TC13 WG2.7/WG13.4 Seventh Working Conference on Engineering for Human-Computer Interaction September 14–18, 1998, Heraklion, Crete, Greece*, ed. by S. Chatty, P. Dewan (Springer US, Boston, MA, 1999), pp. 1–18. https://doi.org/10.1007/978-0-387-35349-4_1
10. H. Garfinkel, Studies of the routine grounds of everyday activities. Soc. Probl. **11**(3), 225–250 (1964). https://doi.org/10.2307/798722
11. J.J. Gibson, *The Ecological Approach to Visual Perception* (Houghton Mifflin, Boston, MA, 1979)
12. J.J. Gibson, Notes on affordances, in *Reasons for realism, Hillsdale* (1982)
13. E. Goffman, *Frame Analysis: An Essay on the Organization of Experience* (Harvard University Press, Cambridge, MA, 1974)
14. J. Han, H. Forbes, D. Schaefer, An exploration of how creativity, functionality, and aesthetics are related in design. Res. Eng. Design **32**(3), 289–307 (2021). https://doi.org/10.1007/s00163-021-00366-9
15. M. Hassenzahl, The thing and I: understanding the relationship between user and product, in *Funology 2: From Usability to Enjoyment* ed. by M. Blythe, A. Monk (Springer International, Cham, Switzerland, 2018), pp. 301–313. https://doi.org/10.1007/978-3-319-68213-6_19
16. H.-M. Escribano, Affordances and social normativity: steps toward an integrative view, in *Affordances in Everyday Life: A Multidisciplinary Collection of Essays*, ed. by Z. Djebbara (Springer International, Cham, Switzerland, 2022), pp. 61–71 . https://doi.org/10.1007/978-3-031-08629-8_7
17. J. Heskett, *Toothpicks and Logos: Design in Everyday Life* (Oxford University Press, New York, NY, 2002)
18. E.L. Hutchins, J.D. Hollan, D.A. Norman, Direct manipulation interfaces. hum.-comput. Interact. **1**(4), 311–338 (1985). https://doi.org/10.1207/s15327051hci0104_2
19. H. Ishii, B. Ullmer, Tangible bits: towards seamless interfaces between people, bits and atoms, in *Proceedings of the ACM SIGCHI Conference on Human Factors in Computing Systems* (ACM, New York, NY, 1997), pp. 234–241
20. R. Jakobson, Linguistics and poetics, in *Style in Language*, ed. by T. Sebeok (The MIT Press, Cambridg,e, MA, 1960), pp.350–377
21. T. Jenkins, I. Bogost, Designing for the internet of things: prototyping material interactions, in *CHI'14 Extended Abstracts on Human Factors in Computing Systems* (ACM, 2014), pp. 731–740. https://doi.org/10.1145/2559206.2578879
22. V. Kaptelinin, B.A. Nardi, *Acting with Technology: Activity Theory and Interaction Design* (MIT Press, Cambridge, MA/London, UK, 2006)
23. K. Kim, H. Lee, M. Tufail, Classification of transformable products based on changes in product form and function. Proc. Des. Soc. **1**, 641–650 (2021). https://doi.org/10.1017/pds.2021.64
24. I. Klimenko, T. Berdnik, Meaning, function and design of object in culture. Postmod. Open./ Deschid. Postmod. **9**(2), 110–119 (2018). https://search.ebscohost.com/login.aspx?direct=true&AuthType=ip,uid&db=asx&AN=130523850&lang=it&site=eds-live&scope=site
25. J. Kolko, *Thoughts on Interaction Design* (Morgan Kaufmann, Burlington, MA, 2010)
26. K. Krippendorff, On the essential contexts of artifacts or on the proposition that "design is making sense (of things)." Des. Issues **5**(2), 9–39 (1989). https://doi.org/10.2307/1511512
27. B. Latour, *Reassembling the Social: An Introduction to Actor-Network-Theory* (Oxford University Press, Oxford, UK, 2005)
28. K.E. MacLean, Haptic interaction design for everyday interfaces. Rev. Hum. Factors Ergon. **4**(1), 149–194 (2008)
29. E. Manzini, *Artefatti. Verso Una Nuova Ecologia Dell'ambiente Artificiale* (Domus Accdemy, Milano, Italy, 1990)
30. E. Manzini, Design in a changing, connected world. Strat. Des. Res. J. **7**(2), 95–99 (2014). https://doi.org/sdrj.2014.72.06
31. E. Manzini, Design culture and dialogic design. Des. Issues **32**(1), 52–59 (2016)

32. I. Mariani, T. Livio, U. Tolino, Drawing interfaces. When interaction becomes situated and variable, in *DeSForM 2019 Proceedings* (DeSForM, Boston, MA, 2019). https://desform19.pubpub.org/pub/drawing-interfaces/release/1
33. M.H. Miraz, M. Ali, P.S. Excell, R. Picking, A review on Internet of Things (IoT), Internet of Everything (IoE) and Internet of Nano Things (IoNT), in *2015 Internet Technologies and Applications (ITA)* (IEEE, 2015), pp. 219–224
34. D.A. Norman, Affordance, conventions, and design. Interactions **6**(3), 38–43 (1999). https://doi.org/10.1145/301153.301168
35. D.A. Norman, S.W. Draper, *User Centered System Design; New Perspectives on Human-Computer Interaction* (Lawrence Erlbaum Associates Inc., Mahwah, NJ, 1986)
36. M.G. Petersen, O.S. Iversen, P.G. Krogh, M. Ludvigsen, Aesthetic interaction: a pragmatist's aesthetics of interactive systems, in *Proceedings of the 5th Conference on Designing Interactive Systems: Processes, Practices, Methods, and Techniques* (Association for Computing Machinery, New York, NY, USA, 2004), pp. 269–276. https://doi.org/10.1145/1013115.1013153
37. E.S. Reed, *Encountering the World: Toward an Ecological Psychology* (Oxford University Press, 1996)
38. C. Rowland, E. Goodman, M. Charlier, A. Light, A. Lui, *Designing Connected Products: UX for the Consumer Internet of Things* (O'Reilly Media Inc., Sebastopol, CA, 2015)
39. M. Segundo-Ortin, Socio-cultural norms in ecological psychology: the education of intention. Phenomenol. Cogn. Sci. (2022). https://doi.org/10.1007/s11097-022-09807-9
40. B.F. Spencer Jr., M.E. Ruiz-Sandoval, N. Kurata, Smart sensing technology: opportunities and challenges. Struct. Control. Health Monit. **11**(4), 349–368 (2004). https://doi.org/10.1002/stc.48
41. T. Sutton, Invisible applications. UX design without GUIs. Mediu. (J. Electron.) (2014). https://medium.com/@thomas_thinks/invisible-applications-10501f6cfa5
42. J. Tidwell, *Designing Interfaces: Patterns for Effective Interaction Design* (O'Reilly Media, Sebastopol, CA, 2010)
43. U. Tolino, I. Mariani, Hacking meanings. Innovation as everyday invention. DIID Disegno Ind. **65**(18), 78–85 (2019)
44. J. Utterback, B. Vedin, E. Alvarez, S. Ekman, S. Walsh Sanderson, B. Tether, R. Verganti, Design-inspired innovation and the design discourse, in *Design-Inspired Innovation* (2006), pp. 154–186
45. R. Withagen, D. Araújo, H.J. de Poel, Inviting affordances and agency. New Ideas Psychol. **45**, 11–18 (2017). https://doi.org/10.1016/j.newideapsych.2016.12.002
46. R. Withagen, H.J. de Poel, D. Araújo, G.-J. Pepping, Affordances can invite behavior: reconsidering the relationship between affordances and agency. New Ideas Psychol. **30**(2), 250–258 (2012). https://doi.org/10.1016/j.newideapsych.2011.12.003
47. A. Yang, S. Rebaudengo, When objects talk back. Medium (2014). https://medium.com/@frogdesign/when-objects-talk-back-8eff3b252482
48. Y. Yoo, Digital materiality and the emergence of an evolutionary science of the artificial, in *Materiality and Organizing: Social Interaction in a Technological World* (2012), pp. 134–154

Chapter 6
Shaping Hidden Interfaces for Situated Interactions

Abstract The chapter investigates the development of hidden and variable interfaces that are responsive to user's interaction and surrounding ecosystems. It presents Think's experimental work in designing interfaces capable of altering an object's functionality, highlighting the adaptability and user-environment interaction. The discourse unfolds through the exploration of *Drawing User Interfaces* as a concept and *Knob!* as its application to a smart controller that changes its interface depending on its position, inclination, or spatial location. The chapter concludes discussing the implications of such variable interfaces for design practices and user experience.

Keywords Variable user interfaces · Visual design · Context-aware technology · Smart ecosystem · Seamless integration

6.1 Towards Context-Aware Interfaces

The technological and computational landscape positions the concept of interaction between objects and the surroundings within a dimension of continuous evolution, especially culturally. Illustrative is how ubiquitous computing [11] and IoT [13] enhance an increasing number of devices across various contexts as networks of interconnected smart elements able to effectively augment our environments, making them computational, interactive, and capable of dialoguing with the surrounding environment and people [26]. As a consequence, there is a growth of interconnected ecosystems where intelligent and sensory objects learn from the context in which they are immersed, adapt to its circumstances, making our interactions increasingly pervasive, dynamic, transparent, tangible, and sometimes extremely natural.

The significant advantage of a progressive increase in computing capacity (ubiquitous computing) lies in reducing the information users need to process and remember during interactions [1], aligning with Sweller's concept of *cognitive load* [33, 34]. This concept refers to the amount of mental effort being used in the working memory, representing a crucial framework in cognitive psychology, particularly relevant in educational and instructional design. It helps in understanding how information can

U. Tolino and I. Mariani, *Design Behind Interaction*,
PoliMI SpringerBriefs, https://doi.org/10.1007/978-3-031-67416-7_6

be presented to optimise learning and comprehension. In the context of ubiquitous computing and IoT, its significance is related to designing systems and interfaces that minimise the cognitive demands on users by reducing unnecessary mental effort. In doing so, user experience can be enhanced, making interactions more intuitive, efficient, and seamless.

Key in the discourse is the development of smart sensors and actuators, which are integrated and networked, leading to various attempts to surpass traditional interfaces—Graphical User Interfaces (GUIs)—and fueling several research lines, including Natural User Interfaces (NUIs) and Tangible User Interfaces (TUIs). Briefly, in NUIs, the logic centres on the absence of a visible interface, i.e., user interfaces that are invisible to their users or become so through successive interactions [2, 17, 24]. Such interfaces open a critical discourse on the necessity to learn how to use them [30]. Despite their intuitive design, the challenge lies in the fact that users are required to adapt to new modes of interaction that are not immediately apparent or familiar. NUIs leverage gestures, voice commands, and other natural interactions, which can be less transparent and more context-dependent. This necessitates a learning curve where users must understand and internalise the nuances of these interactions, which can vary significantly across different devices and contexts.

TUIs involve incorporating the interface into the object using the manipulation logic typical of the analog dimension [19, 32, 40]. This means that users interact with digital information through physical objects and surfaces. By embedding interactive capabilities within everyday objects, TUIs make digital information tangible, and allow users to control digital functions through direct physical manipulation. The approach aims to increase the intuitiveness of digital interactions, leveraging the natural affordances and physical properties of objects, hence lessening the cognitive load required for the interaction.

Recognising the potential of NUIs for their immediate and unmediated interactions, and the appeal of TUIs for their ability to make data tangible, Thingk has explored the development of interfaces focusing on responsiveness and adaptability to context. This includes aspects such as situatedness, context awareness, and context responsiveness. In a landscape where interaction is increasingly pervasive and dynamic, with a growing emphasis on context-sensitive and context-responsive dimensions, Thingk has focused on creating interfaces that adapt to broad user needs [31].

Thingk's approach leverages the concept that varying 'skins' can activate different functionalities, thereby altering users' interpretations and interactions with these objects [15, 23, 26, 37]. Additionally, this reasoning considers how the level of embodiment achieved is such that technology itself can be conceptualised as a design material operating in symbiosis with other technologies and augmented physical materials, in an integrated ecosystem that opens up the experimentation of new experiences and usage practices. Specifically, advances in research and the level of development in the IoT, IoE, and AIoT domains [5, 8, 9, 20, 22, 42] are investigating and seeking interactions where users are not required to input information, but rather networks of intelligent sensors enable computational machines to understand, predict, and interact. This bridge between the physical and digital worlds has become possible thanks to the introduction of sensors, microcontrollers, and actuators, and is fueled by their continuous improvement, along with their economic accessibility and ability to integrate into complex systems. Hidden and variable interfaces leverage a combination of sensors, actuators, and advanced software algorithms to reach a high degree of responsiveness and adaptability. Examples of technology are proximity sensors, gyroscopes, and accelerometers that make these interfaces able to detect user presence, orientation, or movement, and environmental changes, adjusting their functionality accordingly.

These interfaces are designed to impact the objects they are applied to, making them act simultaneously as representations of information and as controllers. The aim is to achieve a unity between interface and interaction, allowing users to interact directly with the information. This approach considers that designing hybrid and intelligent objects, where the physical and digital dimensions merge, necessitates reflection on how these objects are perceived and interpreted by users, as well as how they influence usage habits.

6.2 Drawing User Interfaces

The current technological context and maturity, along with the directions of ongoing experiments and research, and the possibilities arising from including a material perspective, open significant opportunities in the fields of interaction design and HCI. In this context, the knowledge and the multidisciplinary and transdisciplinary perspectives with which we examine the role of materials, environments, and the physical body of the user—as a combination—play a crucial role in interactions involving integrated technologies [7, 16, 21]. The goal is to capture and leverage the principles of digital transformation [25], while simultaneously pursuing technological hybridisation through the fusion of the analog and digital immaterial dimensions.

This subsequent reasoning delves into the concept of Drawing User Interfaces (DUIs) [27, 36], namely interfaces that draw themselves, using sensors to gather information from the physical world (users, the environment, and ecosystems of smart objects), interpret it, and react by altering their aesthetic and functional characteristics [39]. These interfaces dynamically transform in response to necessity and context [32, 38]. This concept is the result of Thingk's experimental research on interfaces capable of modifying an object's functionality by appearing, changing, and disappearing as needed.

6.2.1 (In)visibility and Control

Recognising the limitations and constraints that characterise TUIs [10, 16] and NUI [14, 30], and being aware of the issues arising from the so-called invisibility dilemma

[22], a type of dynamic interface has been conceived, capable of expressing and communicating to the user what its form conceals: its affordances.

In the attempt to integrate intelligent functionalities into everyday objects without falling into the aforementioned dilemma, designers often face a choice between minimising the chances of missing the main task of an interface and adding value by including explicit interactions. When faced with this choice, the tendency is to conceal the object's intelligence, risking making its functions less obvious or evident. However, an increase in functionalities, if not adequately communicated, results in a dichotomy between aesthetics and functions. It becomes crucial to consider how the capabilities introduced into smart objects can be conveyed to the user so that the objects themselves can be appropriately used.

Simultaneously, a second dilemma has been considered, that of control [43], has been considered, being a crucial issue when it comes to smart automation. The focus is on designing objects that act and react autonomously to certain inputs while keeping the underlying complexity invisible to the user. However, this must be achieved in a balanced manner so that the user does not feel deprived of their sense of control over what is happening. A significant case study in this regard is one of the early versions of the *Nest Learning Thermostat*. Users of the thermostat reported feeling uneasy due to not fully understanding the artefact's learning process, which gathers information about their habits to self-regulate the temperature throughout the day. Consequently, users felt unable to fully trust the self-setting of certain functions of the home automation device [44]. The control dilemma requires paying attention to the user, considering their need to be informed about what is happening. Both dilemmas cannot be overlooked when defining a new concept of interface as in the case of DUIs.

Focusing on the functionalities of objects often raises the issue of the amount of information an interface should provide to fulfil each intended activity. The advantage of being able to exploit different sets of variable commands enables to isolate and "situate," or contextualise, the main functions of a device based on the actual use case.

Adopting DUIs allows for designing usage experiences that are situated, variable in time and space, and closely connected to the context in which they are found, reacting with an interface change based on, for example, their positioning, inclination, and user location. DUIs can be thought of as overlapping layers that activate when needed. Each layer is an interface designed to manage specific functionalities.

In this discourse, aesthetic adaptability is a key aspect, leading to the conception of DUIs as interfaces to be integrated into morphological archetypes. The object remains unchanged in its form but changes its communicative skin, offering different functions depending on the interpretation of the surrounding variables, which trigger a fluid and human-centred user experience. Each layer of the object becomes a repository of dynamic information that will be communicated based on its use by the user. Aesthetic adaptability refers to the design feature that enables an interface or object to seamlessly and dynamically alter its visual and functional attributes based on the context of use. This concept emphasises the importance of creating interfaces that are both functional but and visually pleasant in respect to the object they are integrated into. Aesthetically adaptable interfaces maintain the object's integrity and simplicity when not in active use, preserving a clean and minimalistic appearance. However, upon interaction, these interfaces can reveal additional functionalities and information tailored to the current context, providing relevant feedback and controls that enhance user experience. This adaptability ensures that the interface remains unobtrusive and aesthetically pleasing, manifesting its complexity and capabilities as needed, thereby creating a more intuitive and engaging user interaction.

This design scenario increases the variables at play, embedding an *auto-nomatic dimension* [3] that enhances the nature of the elements involved, assigning a performative value to the interface. This dimension refers to a design feature where objects or interfaces have the inherent capacity to act autonomously in response to changes in the environment or to user interactions. This concept stresses the capacity of certain designs to dynamically and autonomously adjust and perform tasks without requiring direct user input for every action. This value generates the ability to act and react at the moment it is manipulated during an interaction [35]. Following this direction, DUIs are designed to automatically recognise and respond to various stimuli or contextual changes, enhancing its interactive and functional capabilities. Moreover, the self-acting feature assigns a performative value to the interface, which becomes able to execute functions, adapt, and react independently, thereby enriching the user experience. A consequence of this characteristic is the need to design considering spatial placement and thus also the visual rhythm, not merely as positional and static identities [4] but as elements that are fluid and dynamic in their composition.

6.2.2 Technology and Experimentation

In light of this, the discourse leverages knowledge from three different applications of DUIs on everyday objects, developed within the context of the H2020 DecoChrom project.[1] Specifically, *Slab! XL*, *Disc! XL*, and *Coin!*. In the project, Think acted as a partner experimenting with technology and developing technological applications in a pilot setting. From an engineering and design perspective, this experimentation tackled the challenge of developing objects that reconfigure their own function according to the interface activated, in a logic of quick adaptability to the context of use [18, 40]. Within this framework, three working prototypes were developed and tested by DecoChrom partners and a community of prosumers. The evaluation process employed an integrated methodology of rapid ethnography, participant observation, questionnaires, and focus groups, all aimed at gathering insightful data

[1] DecoChrom—Decorative Applications for Self-Organized Molecular Electrochromic Systems, was a EU research project funded under the H2020 program, call H2020-EU.2.1.3.—INDUSTRIAL LEADERSHIP – Leadership in enabling and industrial technologies—Advanced materials. GA. 760973, duration 2018-2022. Link: cordis.europa.eu/project/id/760973; decochrom.com.

to pinpoint potential technological and aesthetic improvements. One of the experimentations led to a new version of *Slab!* equipped with an electrochromic display that dynamically visualised its functions, seamlessly integrating the product interface.

To comprehend how these innovative surfaces were perceived and interpreted, an experiment was conducted in a workshop setting, engaging more than 20 lead users to test *Slab! XL* and its redesigned interface, thereby verifying their responses to these novel interactions. Data collection adopted a mixed-method approach, consisting of a qualitative survey (individual), participant observation during individual interactions with the artefact, focus groups, and in-depth interviews. The main findings from the experimentation with *Slab! XL* revealed significant insights into the perception and implications of an adaptive object interface [12] in real-time user interactions, offering a robust foundation for further development and refinement of DUIs. The incorporation of an electrochromic display was generally well-received, with users perceiving it as a calm technology, seamlessly blending into the user experience without being intrusive or distracting. Notably, the feature that stood out was the activation of the electrochromic display through movement rather than a traditional button, which was described by participants in the workshop and focus groups as "surprising", "curious", and "fascinating". However, feedback also highlighted a couple of areas for improvement. Participants noted the relatively low transition speed of the electrochromic display, which could potentially hinder the immediacy of feedback in certain applications. Additionally, the electrochromic ink's inability to achieve complete transparency led to occasional misunderstandings about the display's status, indicating a need for enhanced clarity in its design. These findings underscore the importance of optimising both the transition dynamics and transparency levels of the electrochromic materials to enhance user comprehension and interaction fluidity.

The experimentation we conducted using electrochromic can be implemented by exploring other technological possibilities. The ongoing dematerialisation of physical interfaces in objects has often delegated the operational dimension of product functions to digital screens. Displays embedded within objects or transferred to other devices (such as mobiles and tablets) emphasise the technological dimension and appearance of smart products.

At the current technological stage, smart materials capable of reconfiguring their shape are beginning to emerge, but the creation of a material flexible enough to display a continuous flow of digital data through changes in intrinsic physical properties is still distant. The application field of Drawing User Interfaces (DUI) straddles both the analog (aesthetics and materials) and the digital (functions and transitions) realms. If one opts for the absence of an LED display, changes in the chemical properties of the material (electrochromic) or physical properties (e-ink) can be achieved. Both technologies can be integrated into surfaces made of natural materials—such as wood, marble, and metal—without compromising their appearance and tactile characteristics.

E-ink provides excellent graphic resolution but is limited to a monochromatic range [28]. Among the e-inks, electrochromic screens have the significant advantage of being producible with single-colour screen printing on transparent surfaces, allowing them to reside above the material from which the object is made. Another benefit of this technology is that electrochromics only consume energy during the transition from a transparent state to an opaque one, with no energy consumption required to maintain the state once achieved. This latter solution became the focus of our experimentation within the DecoChrom project, which has overseen its conceptualization and will see its implementation, in prototype form, during upcoming experimental activities.

6.2.3 *Knob!*

The research conducted at Thingk and the technological experimentation within the DecoChrom project nurtured the conceptualisation and design of *Knob!*, an innovative artefact designed to adapt its interface based on the context and specific usage requirements.

Knob! is a DUI concept, embodying interfaces that leverage sensors to collect data from the physical world and respond accordingly. In this framework, the user, environment, and potentially other nearby objects serve as variables that trigger the interface, consequently activating specific functions. This design approach allows *Knob!* to dynamically adjust its displayed information and functionalities based on contextual inputs, creating an adaptive and intuitive user experience (Fig. 6.1).

Specifically, *Knob!* is designed to change its interface according to its positioning, inclination, or location within a space. This versatility is manifested in three distinct use cases:

- **Movement**: *Knob!* responds to being moved or relocated within an environment, altering its interface in response to physical displacement.
- **Reaction**: The object reacts to environmental changes, such as the presence of people, physical variations, or other modifications in or of its context. This adaptability allows it to seamlessly integrate with its surroundings.
- **Dialogue**: *Knob!* can vary its function based on its relationship with other connected objects present in the surrounding space, engaging in a kind of functional dialogue with the surroundings.

Following these three use cases, *Knob!* is a smart controller able to manage different kinds of interactions with the environmental context. As depicted in Fig. 6.1, *Knob!* is a cylindrical object embedded with smart sensors that enable the regulation of various parameters such as temperature, light intensity, and music control. This design can incorporate multiple functional layers, each dedicated to a specific control task. The interface remains inactive, adhering to the principles of calm technology, until it is engaged through interactions like movement, environmental reaction, or dialogue with other devices. Upon interaction, *Knob!* reveals its capabilities. Using the examples of applications presented in Fig. 6.1, one layer could display a rotary interface for temperature adjustment, allowing users to control heating or cooling by rotating the cylinder; another layer might present a linear scale for dimming

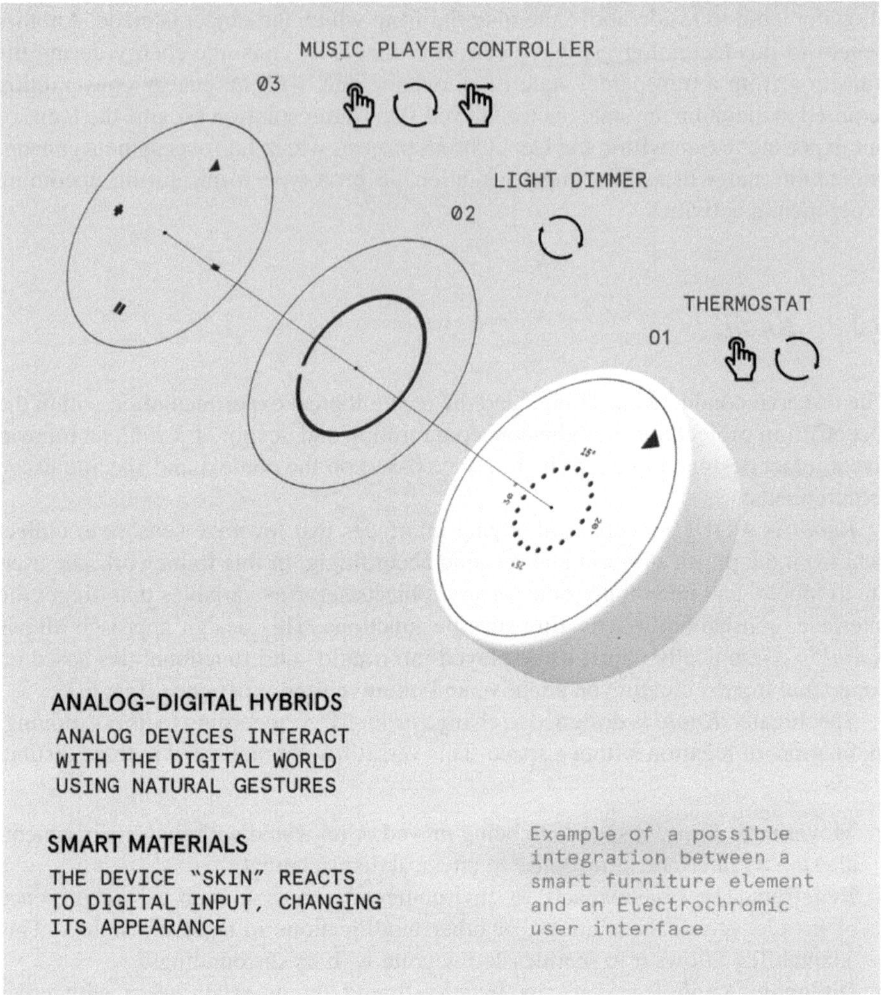

Fig. 6.1 *Knob!* and its multiple interfaces surfacing according to the positioning and interaction with other objects

lights, where users can modify brightness through rotation; a further layer could provide controls for media playback, enabling users to navigate through music or video content with functions such as selection, rotation to skip tracks or chapters, and scrolling or swiping to adjust volume or navigate menus. This design approach makes *Knob!* a dynamic repository of interactive information, seamlessly integrating into various contexts to enhance user experience through intuitive and context-aware interactions.

Through the three use cases discussed, *Knob!* can seamlessly transition between different modes, providing a versatile and user-friendly control mechanism for various smart home applications. These three examples are illustrative of the controller's potential, but the interface can be expanded with a much broader array of functionalities. Furthermore, the interfaces are designed to be updatable and upgradable over time. This flexibility allows for the incorporation of various technologies beyond e-ink, enhancing the controller's adaptability and longevity in response to evolving technological advancements.

This exploration advances reflections on how an object can perform different functions by interpreting surrounding variables, thereby facilitating a fluid and tailored user experience [6]. As circumstances evolve, so does *Knob!*'s communicative skin, effectively conveying its varying functions. The research phase led to the development of functional layers that overlap within a single object, endowing it with multiple capabilities. This approach encourages seamless transitions between various commands and functions that coexist in the same object. The outcome of this investigation is a transitional surface that can smoothly change and react during the interaction.

This case of *Knob!* highlights how designing responsive and adaptable interfaces following an auto-nomaic dimension requires not only that smart objects operate efficiently but also that they resonate intuitively with users. Adhering to this dimension, the device can autonomously detect variations in its positioning, inclination, or proximity to other objects, and adjust its displayed information and functionalities accordingly. Moreover, this approach reduces the cognitive load on the user, since its input for manual adjustments is minimised while enhancing the overall fluidity of the interaction. As a consequence, the system becomes more responsive and adaptive to dynamic and evolving conditions.

6.3 Perspectives: User Experience and Technology

The chapter argues that such interfaces represent a paradigm shift in human-object interaction, moving beyond static designs to more fluid and responsive systems [20]. This shift challenges traditional notions of interface design [29], pushing towards a future where interfaces are not fixed but evolve in real-time, responding to the nuances of user interaction and environmental context.

The approach to interface design described in this chapter demonstrates a design-driven experimentation closely tied to the domain of smart products, with user research and technological trends analysis at its core. Innovation here is focused on the relationship between function and aesthetics related to the context of use.

A significant achievement of this approach is the ability to preserve a minimalist aesthetic while integrating an interactive layer that can become visible or invisible as required. This dynamic layer contains multiple interfaces designed to blend seamlessly with the object's primary aesthetic, ensuring that the product retains its original look. This design strategy allows these interfaces to be embedded in various objects and products, maintaining their visual simplicity while incorporating advanced functionalities that appear only when necessary. In so doing, not only the user experience is enhanced by providing intuitive and context-sensitive interactions but also ensures that the technological enhancements do not detract from the object's visual and tactile appeal.

In this process, it is crucial to understand the role of aesthetics in dynamic digital-physical ecosystems. Designing interfaces for smart, distributed, hyperconnected, and complex ecosystems requires considering the implications of having multiple meanings coexisting in the same object. These meanings that are not persistently displayed but emerge from the surface when needed. Thus, it becomes necessary to investigate how users will perceive, interpret, and interact with objects that adopt interfaces that are simultaneously:

- **Hidden**. As they are not always visible.
- **Variable**. Because they dynamically reconfigure based on environmental factors and user interactions.
- Upgradeable. Designed to have longevity, being changeable, updatable, and enhanceable over time.

Future developments include conducting thorough and extensive user research to gain insights into how these interfaces and the interactions they imply impact the overall user experience. Additionally, advancements in electrochromic technology, and the exploration of other emerging technologies are anticipated to further enhance the concept capabilities. This may extend to the integration of smart materials that are not only reactive and dynamic but also capable of shaping, transforming, and informing through their state changes [41]. Such materials can transform and provide real-time feedback, offering a richer, more interactive experience.

In general, the rationale behind DUIs is significant for various disciplines, from design to HCI, as it explores how interfaces can leverage smart technologies to transform themselves, blending digital and physical elements.

Relevant remain discussing the implications of these variable interfaces for design practices and user experience, in the direction of more intuitive, efficient, and personalised interactions. As such, the insights from this research and experimentation are poised to inform future trajectories of interface design, highlighting the importance of adaptability and context-awareness in the evolving landscape of human–computer interaction.

References

1. K. Ashton, That "Internet of Things" thing. RFID J. (2009). https://www.rfidjournal.com/articles/view?4986
2. S. Ballmer, CES 2010: a transforming trend—the natural user interface (2010). www.huffingtonpost.com/steve-ballmer/ces-2010-a-transforming-t_b_416598.html
3. M. Botta, G.C. Smith, G. Anceschi, *Design dell'informazione: tassonomie per la progettazione di sistemi grafici auto-nomatici* (Valentina Trentini, Trento, Italy, 2006)
4. G. Ceriani, *Il senso del ritmo. Pregnanza e regolazione di un dispositivo fondamentale*, vol. 18 (Meltemi Editore srl, Roma, Italia 2003)

5. B. Dong, Q. Shi, Y. Yang, F. Wen, Z. Zhang, C. Lee, Technology evolution from self-powered sensors to AIoT enabled smart homes. Nano Energy **79**, 105414 (2021). https://doi.org/10.1016/j.nanoen.2020.105414
6. P. Dourish, *Where the Action is: The Foundations of Embodied Interaction* (The MIT Press, Cambridge, MA, 2004)
7. P. Dourish, G. Bell, *Divining a Digital Future: Mess and Mythology in Ubiquitous Computing* (The MIT Press, Cambridge, MA, 2011)
8. D. Evans (2011). The internet of things: how the next evolution of the internet is changing everything. *Cisco Internet Business Solutions Group (IBSG)*, pp. 1–11. www.cisco.com/c/dam/en_us/about/ac79/docs/innov/IoT_IBSG_0411FINAL.pdf
9. D. Evans (2012). The internet of everything: how more relevant and valuable connections will change the world, in *Cisco Internet Business Solutions Group (IBSG)*, pp. 1–9. https://www.cisco.com/c/dam/global/en_my/assets/ciscoinnovate/pdfs/IoE.pdf
10. Y. Fernaeus, J. Tholander, M. Jonsson, Towards a new set of ideals: consequences of the practice turn in tangible interaction, in *TEI'08 Proceedings*. (ACM, 2008), pp. 223–230
11. M. Friedewald, O. Raabe, Ubiquitous computing: an overview of technology impacts. Telematics Inform. **28**(2), 55–65 (2011). https://doi.org/10.1016/j.tele.2010.09.001
12. J.J. Gibson, *The Ecological Approach to Visual Perception* (Houghton Mifflin, Boston, MA, 1979)
13. J. Gubbi, R. Buyya, S. Marusic, M. Palaniswami, Internet of Things (IoT): a vision, architectural elements, and future directions. Incl. Spec. Sect.: Cyber-Enabled Distrib. Comput. Ubiquitous Cloud Netw. Serv. Cloud Comput. Sci. Appl. Big Data Scalable Anal. Beyond **29**(7), 1645–1660 (2013). https://doi.org/10.1016/j.future.2013.01.010
14. L.K. Hansen, P. Dalsgaard, Note to self: stop calling interfaces "natural", in *Proceedings of The Fifth Decennial Aarhus Conference on Critical Alternatives* (2015), pp. 65–68
15. J. Heskett, *Toothpicks and Logos: Design in Everyday Life* (Oxford University Press, New York, NY, 2002)
16. E. Hornecker, J. Buur, Getting a grip on tangible interaction: a framework on physical space and social interaction, in *Proceedings of the SIGCHI conference on Human Factors in computing systems* (ACM, 2006), pp. 437–446
17. H. Ishii, Tangible user interfaces, in *Human-Computer Interaction: Design Issues, Solutions, and Applications* (2007), pp. 141–157
18. H. Ishii, The tangible user interface and its evolution. Commun. ACM **51**(6), 32–36 (2008). https://doi.org/10.1145/1349026.1349034
19. R.J. Jacob, A. Girouard, L.M. Hirshfield, M.S. Horn, O. Shaer, E.T. Solovey, J. Zigelbaum, Reality-based interaction: a framework for post-WIMP interfaces, in *Proceedings of the SIGCHI Conference on Human Factors in Computing Systems* (ACM, 2008), pp. 201–210
20. T. Jenkins, I. Bogost, Designing for the internet of things: prototyping material interactions, in *CHI'14 Extended Abstracts on Human Factors in Computing Systems* (ACM, 2014), pp. 731–740. https://doi.org/10.1145/2559206.2578879
21. S.R. Klemmer, B. Hartmann, L. Takayama, How bodies matter: five themes for interaction design, in *Proceedings of the 6th Conference on Designing Interactive Systems* (ACM, 2006), pp. 140–149
22. M. Kranz, P. Holleis, A. Schmidt, Embedded interaction: interacting with the internet of things. IEEE Internet Comput. **2**, 46–53 (2009). https://doi.org/10.1109/MIC.2009.141
23. K. Krippendorff, On the essential contexts of artifacts or on the proposition that "design is making sense (of Things)." Des. Issues **5**(2), 9–39 (1989). https://doi.org/10.2307/1511512
24. G. Krishna, *The Best Interface is No Interface: The Simple Path to Brilliant Technology* (New Riders, San Francisco, CA, 2015)
25. M. Kuniavsky, *Smart Things, Ubiquitous Computing User Experience Design* (Elsevier, Burlington, MA, 2010)
26. B. Latour, On interobjectivity. Mind Cult. Act. **3**(4), 228–245 (1996)
27. I. Mariani, T. Livio, U. Tolino, Drawing interfaces. When interaction becomes situated and variable, in *DeSForM 2019 Proceedings* (eSForM, Boston, MA, 2019). https://desform19.pubpub.org/pub/drawing-interfaces/release/1

28. T. Moreira, M. Maia, A.J. Parola, M. Zangoli, Di Maria, F., C.A.T. Laia, Ink-jet-printed semiconductor electrochromic nanoparticles: development and applications in electrochromism, in *Chemical Solution Synthesis for Materials Design and Thin Film Device Applications*, ed. by S. Das, S. Dhara (Elsevier, 2021), pp. 407–437. https://doi.org/10.1016/B978-0-12-819718-9.00021-2
29. D.A. Norman, Affordance, conventions, and design. Interactions **6**(3), 38–43 (1999). https://doi.org/10.1145/301153.301168
30. D.A. Norman, Natural user interfaces are not natural. Interactions **17**(3), 6–10 (2010). https://doi.org/10.1145/1744161.1744163
31. D. Saffer, *Designing for Interaction: Creating Innovative Applications and Devices* (Peachpit press, Berkeley, CA, 2010)
32. O. Shaer, E. Hornecker, et al., Tangible user interfaces: past, present, and future directions. Found. Trends Human Comput. Interact. **3**(1–2), 4–137 (2010).
33. J. Sweller, Cognitive load theory, learning difficulty, and instructional design. Learn. Instr. **4**(4), 295–312 (1994)
34. J. Sweller, Cognitive load theory, in *Psychology of Learning and Motivation*, vol. 55 (Elsevier, Burlington, MA, 2011), pp. 37–76.
35. U. Tolino, Figure dinamiche e di interazione, in *Progetto e culture visive. Elementi per il design della comunicazione*, ed. by V. Bucchetti (FrancoAngeli, Milano, Italy, 2018)
36. U. Tolino, I. Mariani, T. Livio, S. Marangoni, Variable and situated user interfaces: assumptions, potentials and design issues, in *Proceedings of the 18th International Conference on Mobile and Ubiquitous Multimedia* (Association for Computing Machinery, New York, NY, USA, 2019). https://doi.org/10.1145/3365610.3368424
37. R. Verganti, *Design-Driven Innovation - Changing the Rules of Competition by Radically Innovating What Things Mean* (Harvard Business Press, Boston, MA, 2009)
38. O. Vermesan, P. Friess, *Internet of Things: Converging Technologies for Smart Environments and Integrated Ecosystems* (River Publishers, Delft, The Netherlands, 2013)
39. M. Wiberg, H. Ishii, P. Dourish, D. Rosner, A. Vallgårda, P. Sundström, et al., Material interactions: from atoms & bits to entangled practices, in *CHI'12 Extended Abstracts on Human Factors in Computing Systems* (ACM, 2012), pp. 1147–1150
40. D. Wigdor, D. Wixon, *Brave NUI World: Designing Natural User Interfaces for Touch and Gesture* (Elsevier, Burlington, MA, 2011)
41. F.R. Wilson, *The Hand: How its Use Shapes the Brain, Language, and Human Culture* (Vintage, New York, NY, 1999)
42. J. Yan, AIoT in smart homes: challenges, strategic solutions, and future directions. Highlights Sci. Eng. Technol. **87**, 59–65 (2024). https://doi.org/10.54097/8hzgaf51
43. A. Yang, S. Rebaudengo, When objects talk back. Medium (2014). https://medium.com/@frogdesign/when-objects-talk-back-8eff3b252482
44. R. Yang, M.W. Newman, Learning from a learning thermostat: lessons for intelligent systems for the home, in *Proceedings of the 2013 ACM International Joint Conference on Pervasive and Ubiquitous Computing* (ACM, New York, NY, USA, 2013), pp. 93–102. https://doi.org/10.1145/2493432.2493489

Chapter 7
Conclusions and Future Research

Abstract This chapter outlines the main insights from eight years of research and experimentation at Thingk, highlighting the impact of digital technology on design, human–computer interaction, and user experience. It critically discusses how the development of embedded technology, smart objects, and interactive interfaces led to novel interaction paradigms and design methodologies that bridge theoretical research and market applications.

Keywords Design methodologies · Contribution to knowledge · Research and experimentation · Future research directions · Study limits

7.1 Making Sense of Things

This book offers a critical synthesis of discussions across its chapters, focusing on major contributions to the field of interaction design and technology-enhanced products. It targets researchers, designers, practitioners, and entrepreneurs, providing insights that challenge conventional approaches and suggest new directions for innovation. The text is the result of years of research, presenting not just reflections on smart product design, but a deeper inquiry into the integration of technology within everyday objects. The intent from the outset was to offer a detailed account based on extensive field experiments and a solid theoretical foundation.

Reflecting on years of research and design experimentation, this book captures Thingk's journey through the innovative use of embedded technology, smart objects, and interactive interfaces, exploring how the spin-off continued to an ongoing discourse with the development of new interaction paradigms and design methodologies that effectively bridge theoretical research with practical market applications. Each chapter draws from a rich theoretical base to inform practical case studies, applied through qualitative and quantitative research that underpins the practical outcomes presented. This comprehensive exploration is backed by a multitude of studies, iterative designs, and extensive validations, enriched through interdisciplinary dialogues and collaborations. Specific attention is given to discuss key

U. Tolino and I. Mariani, *Design Behind Interaction*,
PoliMI SpringerBriefs, https://doi.org/10.1007/978-3-031-67416-7_7

findings and implications, revealing the ways objects, designed to keep their interaction modalities hidden until required, reshape user perceptions of agency and affordances. This shows how Thingk's approach not only challenges conventional design paradigms but redefines the interaction ecosystem, underlining the importance of context-aware and variable interfaces for intuitive and personalised user experiences.

7.2 Insights on Operationalising Theories

The academic research conducted significantly informed the development of Thingk's products, leveraging insights and theoretical frameworks established in the study of IoT and smart home technologies.

The products developed by Thingk exemplify the benefits derived from its interdisciplinary approach, integrating diverse fields such as design and engineering, sociology, and cognitive psychology. This melding of disciplines has not only enhanced understanding of user interaction with smart technologies [2, 6, 11], but also spurred the creation of innovative practices. This foundation has enabled Thingk to design products that are not only technologically advanced but also user-friendly and aware of their contexts, ensuring seamless integration into daily environments.

Thingk's perspective has been influenced by the concept of calm technology, which advocates for technology that seamlessly integrates into the user's environment and becomes apparent only when necessary [8, 12]. This is exemplified in *Slab!*, which conceals its digital functionalities within a natural wooden surface, revealing its capabilities only during interaction, thus embodying the principles of calm technology.

This approach enhances usability while maintaining an intuitive user experience, as suggested by research emphasising the importance of technology integration that aligns with an object's functionality and user needs [4]. Thingk applied this by embedding hidden digital displays and touch sensors within natural materials, ensuring that the technological elements enhance rather than detract from the user experience.

Moreover, Thingk prioritises user-centric design principles, focusing on natural and intuitive interactions [1, 3, 7], while being surprising and unexpected. This is evident in *Desk!*, where the wireless charging zones are only activated upon contact with a smartphone, simplifying the user interaction process and enhancing functionality; similarly, Slab! switches between being a kitchen scale and a timer based on its orientation, demonstrating the application of adaptive interaction concepts. These designs showcase adaptive interaction concepts by adjusting functionality in response to user actions and contextual changes [5].

Additionally, the aesthetic and material choices in Thingk's products, such as the use of high-quality wood and leather, which not only provide an elegant tactile experience but also align with theoretical discussions on the importance of materiality in smart product design [9, 10]. This holistic approach to design-driven innovation and user-centric strategies is evident throughout Thingk's methodology, which leverages extensive user research and co-design practices to challenge conventional paradigms and enhance user and stakeholder engagement, ensuring that every product is both functional and aesthetically pleasing while being finely attuned to user feedback and evolving usage contexts.

7.3 Main Contributions to Research

The academic research conducted on IoT and smart technologies has provided a robust theoretical foundation that guided Thingk in developing innovative products. By integrating the principles recalled above, Thingk has successfully translated academic insights into practical, market-ready solutions. This seamless integration of theory and practice exemplifies how academic research can drive real-world innovation in the domain of smart home appliances. In light of this, a significant contribution from Thingk are insights on the operationalisation of theoretical principles and their practical application to product design, illustrating how abstract concepts are transformed into tangible innovations.

Below a summary of how this book contributed to knowledge:

- **Bridging the gap between theory and practice.** By using the case study of Thingk, a university spin-off, the book demonstrates how theoretical concepts are operationalised in real-world settings, offering valuable insights for both academics and practitioners.
- **Interdisciplinary approach.** The research and innovation described builds upon the intreation of knowledge from Interaction, Product, and Communication Design, discourse, reaching out to insights from sociology, anthropology, ethnography, and marketing. This multiplicity is woven into the design process, showcasing a comprehensive approach to product development that intertwines design-driven innovation and technology.
- **Empirical research and design experimentation.** Discussing about Thingk's empirical research and design experimentation, the book provides a comprehensive and evidence-based analysis of design processes and methodologies, contributing to the field of research-through-design.
- **Embedded technology and implications in interaction design.** The book explores how emerging technologies and embedded interfaces are reshaping the aesthetics, functionality, and user interaction of products. This contributes to the evolving discourse on design paradigms, particularly in the context of smart and IoT-enabled objects. The book delves into the development of variable interfaces that are responsive to user's interaction, the surrounding ecosystem, and environmental conditions. This represents a significant shift from static to dynamic design models, contributing to the understanding of how interfaces can evolve to enhance user experience and engagement.
- **Exploration of semantic reconfiguration.** The book focuses on semantic reconfiguration in product design, exploring how objects with hidden or misleading interfaces impact user perception and interaction. This discussion is particularly

relevant in understanding the effects of non-explicit interactions on agency and affordance.

The perspectives offered in this book serve as a valuable resource for researchers and scholars exploring the nuanced interplay of design and technology, as well as a guide for practitioners and designers seeking to deepen their understanding of this dynamic field. For entrepreneurs, the book acts as a knowledge repository, highlighting the real-world applications and implications of designing smart, interactive products.

7.4 Limitations

Throughout this research, we faced several limitations and deviations that provided critical insights into the complex relationship between design aims, technological constraints, and user interactions. These challenges were instrumental in identifying areas ripe for future exploration and emphasised the need for ongoing adaptation in research methodologies. A primary challenge was aligning theoretical frameworks with the practical demands of product development, which underscored the tensions between idealistic concepts and practical implementation. The challenges faced during the integration of technology into user-friendly designs highlighted the fragile balance required between functionality and aesthetic simplicity. These challenges were not merely obstacles but also opportunities for learning and adaptation, prompting re-evaluation of approaches and methodologies. The deviations from the planned research path, therefore, were not just limitations but also enriching elements that broadened the scope of investigation.

Moreover, the integration of cutting-edge materials, such as electrochromic inks, exposed technological gaps between current capabilities and future directions. These technological limitations were not seen as failures but as valuable insights that delineate the current boundaries of the field and pave the way for future advancements. Such deviations from the planned research path enriched our investigation, broadening the scope and deepening the understanding of dynamic interface design within smart products.

7.5 Future Directions

Building on the knowledge and experience gained, relevant future research includes in-depth studies into the behavioural impacts of hidden/drawing interfaces, exploration of emerging technologies in interface design, and expansion into new application domains. Expansion into new application domains could further test the scalability and adaptability of the current findings. Additionally, an ongoing revision of technological capabilities and user expectations will be crucial in maintaining the relevance and applicability of design solutions in a rapidly changing digital landscape.

One of the most promising avenues for future research within the smart home domain is the incorporation of AI, for instance as AIoT. This integration signifies a profound shift towards smarter and autonomous systems capable of enhancing everyday interactions. AIoT brings together AI's learning and decision-making capabilities with IoT's connectivity and data generation and exploitation. Such systems can learn from user behaviour, anticipate needs, and optimise the environment for energy efficiency, security, and comfort.

Additionally, the adoption of AI and other emerging technologies are introducing significant challenges that must be carefully considered and managed. Issues such as technology opacity—the lack of transparency about how smart products operate and make decisions—raise concerns about user trust and understanding. Furthermore, there are critical considerations regarding data usage, particularly how data are collected, processed, and utilised, raising privacy and security concerns. Additionally, there is the risk of excessive reliance on technology, where users may over-trust automated systems to manage home environments without sufficient oversight or understanding.

Despite these challenges, AIoT holds the potential to transform smart homes by making them more responsive and attuned to the needs of their inhabitants. The integration of AI can lead to the development of interfaces that are not only reactive but also proactive, adjusting to changes in the environment or user behaviour before they even enter conscious awareness. This could mean smarter energy management, enhanced security systems that predict potential threats, or health monitoring systems that adjust conditions to support individual wellness proactively. As we navigate these advancements, it is crucial to balance innovation with ethical considerations and user-centric design to ensure that smart homes remain safe, secure, and aligned with the needs of those who inhabit them.

As the adoption of smart home technologies expands, ensuring their accessibility becomes increasingly crucial. The broadening audience for these technologies includes not only tech-savvy users but also elderly, individuals with imparities, and not traditionally tech-oriented persons. This demographic diversification implies that smart home devices be designed with more attention to universal design principles in mind, aiming to create solutions that are usable and beneficial, regardless of age or ability. Moreover, cost considerations are crucial: as smart home technologies proliferate, it becomes relevant not to widen the digital divide but instead reason towards standardised uses, accessible to all households.

Ultimately, this overall evolving landscape requires a new form of literacy that goes beyond basic digital skills, including understanding of the implications of making our houses smarter. For instance, including transparent communication from manufacturers about how devices operate, what data they collect, and how it is used can help build trust and confidence among users, ensuring they are informed and in control of their smart environments.

As smart technology becomes increasingly integrated into our homes, it becomes relevant that users are equipped with the knowledge to comprehend, manage and interact with these systems effectively, consciously, and conscientiously. This new literacy extends beyond merely operating devices; it requires an understanding of data privacy, security implications, and the ethical use of smart technology. Ethical considerations should indeed steer the design and deployment of these technologies to ensure they respect user privacy, offer transparency, and promote fairness in their functionality and data handling.

References

1. L. Gonçalves, L. Patrício, J. Grenha Teixeira, N.V. Wünderlich, Understanding the customer experience with smart services. J. Serv. Manag. **31**(4), 723–744 (2020). https://doi.org/10.1108/JOSM-11-2019-0349
2. H. Kang, K.J. Kim, Feeling connected to smart objects? A moderated mediation model of locus of agency, anthropomorphism, and sense of connectedness. Int. J. Hum Comput Stud. **133**, 45–55 (2020). https://doi.org/10.1016/j.ijhcs.2019.09.002
3. M.J. Kim, M.E. Cho, H.J. Jun, Developing design solutions for smart homes through user-centered scenarios. Front. Psychol. **11**, 335 (2020). https://doi.org/10.3389/fpsyg.2020.00335
4. M. Kranz, P. Holleis, A. Schmidt, Embedded interaction: interacting with the internet of things. IEEE Internet Comput. **2**, 46–53 (2009). https://doi.org/10.1109/MIC.2009.141
5. G. Krishna, *The Best Interface is no Interface: The Simple Path to Brilliant Technology* (New Riders, San Francisco, CA, 2015)
6. D. Marikyan, S. Papagiannidis, E. Alamanos, Cognitive dissonance in technology adoption: a study of smart home users. Inf. Syst. Front. **25**(3), 1101–1123 (2023). https://doi.org/10.1007/s10796-020-10042-3
7. J. Tidwell, *Designing Interfaces: Patterns for Effective Interaction Design* (O'Reilly Media, Sebastopol, CA, 2010)
8. M. Weiser, J.S. Brown, Designing calm technology. PowerGrid J. **1**(1), 75–85 (1996)
9. M. Wiberg, H. Ishii, P. Dourish, D. Rosner, A. Vallgårda, P. Sundström, et al., Material interactions: from atoms & bits to entangled practices, in *CHI'12 Extended Abstracts on Human Factors in Computing Systems* (ACM , 2012), pp. 1147–1150)
10. M. Wiberg, E. Robles, Computational compositions: aesthetics, materials, and interaction design. Int. J. Des. **4**(2), 65–76 (2010)
11. C. Wilson, T. Hargreaves, R. Hauxwell-Baldwin, Smart homes and their users: a systematic analysis and key challenges. Pers. Ubiquit. Comput. **19**(2), 463–476 (2015). https://doi.org/10.1007/s00779-014-0813-0
12. E. R. B. de Wolf, Pervasive technology as scaffolds for mindful living: reframing calm technology (2021). http://essay.utwente.nl/88238/